Dams and Irrigation

By

Dr. M. Lakshmi Narasaiah

M.A., Ph.D.

Professor of Economics,
Coordinator, Department of M.B.A.
Sri Krishnadevaraya University Post-graduate Centre,
Kurnool–518 002
Andhra Pradesh (India)

DISCOVERY PUBLISHING HOUSE
NEW DELHI

First Published–2005
Reprinted-2010

ISBN: 81-7141-978-X

Published by:

DISCOVERY PUBLISHING HOUSE

4831/24, Prahlad Street, Ansari Road, Darya Ganj
New Delhi–110 002 (India)
Phone: 23279245, • Fax: 91-11-23253475
e-mail: dphtemp@indiatimes.com

Printed at:
Sachin Printers
Delhi

Preface

Today, around 3800 km^3 of fresh water is withdrawn annually from the world's lakes, rivers and aquifers. This is twice the volume extracted 50 years ago. World population has passed 6 billion. Projections say that it will reach a peak of between 7.3 billion and 10.7 billion around 2050 before total population begins to stabilise or fall.

50 litres per person per day (or just over 18.25 m^3 a year) covers basic human water requirements for drinking, sanitation, bathing and food preparation. In 1990, over a billion people had access to less than 50 litres of water a day. Agriculture accounts for about 67 per cent of withdrawals, industry uses 19 per cent and municipal and domestic uses account for 9 per cent.

One third of the countries in water-stressed regions of the world are expected to face severe water shortages this century. By 2025 there will be approximately 6.5 times as many people—a total of 3.5 billion living in water-stressed countries. By the end of the 20th century, there were over 45,000 large dams in over 150 countries. The average large dam today is about 35 years old. Since average construction periods generally range from 5 to 10 years, this indicates a worldwide annual average of some 160 to 320 new large dams per year.

During the 1990s, an estimated $32-46 billion was spent annually on large dams, four-fifths of it in developing countries. Of the $22-31 billion invested in dams each year in developing countries, about four-fifths was financed directly by the public sector. About one-fifth of the world's agricultural land is irrigated, and irrigated agriculture accounts for about 40% of the world's agricultural production.

Half of the world's large dams were built exclusively or primarily for irrigation, and an estimated 30 to 40 per cent of the 271 million hectares of irrigated lands worldwide rely on dams. Dams are estimated to contribute to 12-16 per cent of world food production. Hydropower currently provides 19 per cent of the world's total electricity supply, and is used in over 150 countries with 24 of these countries depending on it for 90 per cent of their supply.

Floods affected the lives, on average, of 65 million people between 1972 and 1996, more than any other type of disaster, including war, drought and famine. There are 261 water sheets that cross the political boundaries of two or more countries. A number of key international rivers lack a basin-wide agreement that defines a process for establishing equitable water use between riparian States.

Cost Effectiveness

Cost performance data confirms that large dam projects often incur substantial capital cost overruns. The average overrun was half again as much as the projected cost. The bulk of hydropower projects have delivered power within a close range of pre-project targets but with an overall tendency to fall short of targets.

At current rates, water fees are rarely sufficient to recover both capital and recurrent costs for water supply systems in many developing countries. Growing concern over the cost and effectiveness of large dams and related structural measures as long-term responses to floods has led to support for integrated flood management as opposed to flood control.

Multi-purpose schemes are inherently more complex and many experience operational conflicts that contribute to under-performance on financial and economic targets. Substantive evaluations of project performance are few in number, narrow in scope, and poorly integrated across impact categories and scales.

Dr. M. Lakshmi Narasaiah

Contents

1

Big Dam Construction is on the Rise

As the era of big dams fades in North America, construction is increasing in Asia, fuelled by growing demand for electricity and irrigation. Worldwide, the number of dams under construction in 1998 (the latest year for which global data are available) rose by more than 9 per cent, to at least 1,240, after a much smaller increase in 1992. The accompanying chart shows that China accounts for more than one fourth of the big dams under construction, while China, Japan, South Korea and India together account for more than half. The increases follow a decline in the 1980s, when construction world wide averaged less than half that of the preceding 25 years.

Data for the early 1990s, though incomplete, indicate a shift toward larger dams, with concomitant greater social and economic costs. In 1992, 60 per cent of the dams being built were more than 30 metres high compared with only 21 per cent of existing dams in 1986. Construction of dams higher than 100 metres rose by some 27 per cent between 1991 and 1993, half of these large structures were built by just three countries—Japan, China and Turkey.

Dams are a symbol of modernity and a source of national prestige, partly because they are a multipurpose tool of development. They generated more than 18 per cent of the World's electricity in 1998, and reservoir water irrigates millions of hectares of land around the world, raising agricultural yields two to three times over those of dry land farming, and in many cases bringing agriculture to regions that could not otherwise support it. Areas with a high percentage of irrigated agriculture, such as China, India and the western United States, rely heavily on reservoirs for water. Irrigation is increasingly listed as the primary purpose of new dams, possibly reflecting growing food needs and water scarcity in many developing countries.

Dam reservoirs also protect societies against drought. Water in the world's reservoirs effectively increases the normal supply from rivers by some 30 per cent. This insurance can be invaluable: Egyptians maintain that the Aswan Dam saved Egypt from catastrophe during the drought of 1979-80.

In the past two decades, however, as it became clear that dams have serious costs as well as benefits, dam construction began to decline. Loss of land to reservoirs can be significant. The Narmada Sagar Project in India is expected to drive 31 species of plant to extinction because of habitat changes caused by reservoir creation. Decomposing plant life at reservoir bottoms can release as much methane and carbon dioxide both greenhouse gasses—as a coalfired plant with the same electricity generating capacity, according to the Freshwater Institute in Canada.

Dams also restrict stream flow with often disastrous biological and economic consequences. Populations and fisheries have been nearly eliminated in many places worldwide, as on the Columbia River in the United States, where once plentiful salmon are now endangered species, and valuable fisheries have collapsed. Changes in temperature and water alter habitats of fish and other wildlife. Sediment that would normally enrich flood plain soils gets trapped behind dams. This last problem, siltation, can shorten a dam's useful life; in an extreme case, the turbines of China's Sanmenxia Dam were shut down in 1964 after only four years of operation when the dam's reservoir filled with sediment.

By restricting stream flows, dams can also contribute to the decline of local economies. One example is Indian diversion of the Ganges river before it reaches Bangladesh, which has caused $25 million in losses for an agricultural project in northwest Bangladesh, and increased salination of fishing areas at the Bay of Bengal, the river's natural outlet.

Dams exact a high human toll as well. Millions of people were resettled in the past half century to make room for dam reservoirs. China's Three Gorges Dam alone may well displace more than one million people. Those affected have had little say in their resettlement, and most displaced citizens wind up in worse situations than before their move.

Dams can increase health risks. Waterborne disease, including malaria and river blindness, have been introduced and

TABLE
LEADING BUILDERS OF BIG DAMS
(HIGHER THAN 10 METRES)[1]

Country	New Starts[2] (Number)	Under Construction (Number)
China	82	311
Turkey	84	190
Japan	11	140
South Korea	2	125
India	48	76
United States	30	55
Spain	16	53
Romania	0	39
Italy	0	37
Tunisia	16	28
Algeria	6	27
Iran	1	76
Thailand	7	17
Greece	3	14
France	8	12
Brazil	4	12

spread as reservoirs are created. Large-scale water projects were a major contributor to the 75 per cent global increase in cases of schistosomiasis, sometimes fatal disease.

Advocates of dams acknowledge these shortcomings, but argue that these effects can be mitigated, or that the alternatives to dams would be worse. Others argue that evaluation of dams involves widely differing environmental, social and political contexts, and a project-by-project analysis of each dam is needed.

The World Bank—long a strong supporter of big dams—appears to have slowed its involvement. The Bank was involved with an average of 18 dam projects a year between 1980 and 1985, but with only 6 a year between 1986 and 1993. In fact, the Bank established a commission in 1994 to hear complaints from parties affected by proposed dam projects, the first of whom were people affected by the Arun III dam in Nepal.

The future of dams is unclear. Proponents note that only a fraction of the technically usable hydropower potential in developing countries has been harnessed. But the discussion has moved beyond technical considerations. Wide ranging development, conservation, energy use and population considerations are or should be part of every proposed dam's calculus. It remains to be seen whether current and future big dams can meet the increasingly stringent environmental and social standards expected of them, and whether development needs can be met in other ways.

2

Water Facts and Findings on Large Dams

Water Facts

Today, around 3800 km^3 of fresh water is withdrawn annually from the world's lakes, rivers and aquifers. This is twice the volume extracted 50 years ago. World population has passed 6 billion. Projections say that it will reach a peak of between 7.3 billion and 10.7 billion around 2050 before total population begins to stabilise or fall.

50 litres per person per day (or just over 18.25 m^3 a year) covers basic human water requirements for drinking, sanitation, bathing and food preparation. In 1990, over a billion people had access to less than 50 litres of water a day. Agriculture accounts for about 67 per cent of withdrawals, industry uses 19 per cent and municipal and domestic uses account for 9 per cent.

One third of the countries in water-stressed regions of the world are expected to face severe water shortages this century. By 2025 there will be approximately 6.5 times as many people— a total of 3.5 billion living in water-stressed countries. By the end of the 20th century, there were over 45,000 large dams in over 150 countries. The average large dam today is about 35 years old. Since average construction periods generally range from 5 to 10 years, this indicates a worldwide annual average of some 160 to 320 new large dams per year.

During the 1990s, an estimated $32-46 billion was spent annually on large dams, four-fifths of it in developing countries. Of the $22-31 billion invested in dams each year in developing countries, about four-fifths was financed directly by the public sector. About one-fifth of the world's agricultural land is irrigated, and irrigated agriculture accounts for about 40 per cent of the world's agricultural production.

Half of the world's large dams were built exclusively or primarily for irrigation, and an estimated 30 to 40 per cent of the 271 million hectares of irrigated lands worldwide rely on dams. Dams are estimated to contribute to 12-16 per cent of world food production. Hydropower currently provides 19 per cent of the world's total electricity supply, and is used in over 150 countries with 24 of these countries depending on it for 90 per cent of their supply.

Floods affected the lives, on average, of 65 million people between 1972 and 1996, more than any other type of disaster, including war, drought and famine. There are 261 water sheets that cross the political boundaries of two or more countries. A number of key international rivers lack a basin-wide agreement that defines a process for establishing equitable water use between riparian States.

Cost Effectiveness

Cost performance data confirms that large dam projects often incur substantial capital cost overruns. The average overrun was half again as much as the projected cost. The bulk of hydropower projects have delivered power within a close range of pre-project targets but with an overall tendency to fall short of targets.

At current rates, water fees are rarely sufficient to recover both capital and recurrent costs for water supply systems in many developing countries. Growing concern over the cost and effectiveness of large dams and related structural measures as long-term responses to floods has led to support for integrated flood management as opposed to flood control.

Multi-purpose schemes are inherently more complex and many experience operational conflicts that contribute to under-performance on financial and economic targets. Substantive evaluations of project performance are few in number, narrow in scope, and poorly integrated across impact categories and scales.

Ecological Costs

Dams, inter basin transfers, and water withdrawals for irrigation have fragmented 60 per cent of the world's rivers. As a physical barrier the dam disrupts the movement in upstream and downstream species composition and even species loss. In Africa, the changed hydrological regime of rivers has adversely affected floodplain agriculture, fisheries, pasture and forests that constituted the organising element of community livelihood and culture.

Problems may be magnified as more large dams are added to a river system, resulting in an increased and cumulative loss of natural resources, habitat quality, environmental sustainability and ecosystem integrity. Good site selection, such as not building large dams on the main-stem of a river system, and better dam design also played significant roles in avoiding or minimising impacts. The economic appraisal techniques such as risk and distributional analysis were still mandated for only 20% of large dam projects even in the 1990s.

Social Costs

At the planning and design stage, an important social impact is the delay between the decision to build a dam and the onset of construction. This can result in communities living for decades starved of development and welfare investments.

The overall global level of physical displacement could range from 40 to 80 million. In India and China together, large dams could have displaced between 26-58 million people between 1950 and 1990. Little or no meaningful participation of affected people in the planning and implementation of dam projects—including resettlement and rehabilitation has taken place.

Empowering people, particularly the economically and socially marginalised by respecting their rights and ensuring that resettlement with development becomes a process governed by negotiated agreements is critical to positive resettlement and rehabilitation.

Poor accounting in economic terms for the social and environmental costs and benefits of large dams implies that the true economic efficiency and profitability of these schemes remains largely unknown.

The direct adverse impacts of dams have fallen disproportionately on rural dwellers, subsistence farmers, indigenous people, ethnic minorities, and women. Where costs and benefits accrue to different groups, the standard procedures for adding up and discounting the expected costs and benefits do not provide an appropriate measure of changes in social welfare.

Financing

Almost 2 billion people, both urban and rural poor, have no access to electricity at all. Most efficiency measures and technologies are cost-effective at today's electricity prices and the use of full environmental and social costing of electricity supply options makes them even more so. Among advanced technologies in research and development, micro turbines and fuel cell show the greatest near and mid-term promise.

Total financing for large dams from multi-lateral and bi-lateral development banks comes to more than $4 billion annually at the peak of lending during 1975-84. Although the proportion of investment in dams directly financed by bi-laterals and multilateral was perhaps less than 15%. The total investment in dams by the multilaterals and bilaterals since 1950 is approximately $125 billion.

3

A Breakthrough in the Evolution of Large Dams

Back to the Negotiating Table

"The problem is not the dams." It is the hunger. It is the thirst. It is the darkness in a township." With these plain words, former South African President Nelson Mandela summed up the World Commission on Dams (WCD's) motives in a speech at the presentation praising its work. His position is similar to that of the many other representatives of the South who were present: he demands a right to development. But Ms Medha Patkar, of India, a WCD Commission member and founder of the struggle to save the Narmada River (Narmada Bachao Andolan) anti-dam movement, takes a contrary view. "The problems of the dams are only a symptom of the larger failure of the unjust and destructive dominant development model," she says. We need to challenge "the forces that lead to the marginalisation of a majority through the imposition of unjust technologies like large dams."

Stagnation in Dam Building

The WCD is a unique experiment in reaching consensus. It began in April 1997 when, with the support of the World Bank and the World Conversation Union (IUCN), 39 representatives of diverse interests met at a workshop in Gland, Switzerland. At this point the various positions of the participants from governments, the private sector, international financial institutions, civil society organisations and affected people were cast in stone. The Manibeli Declaration in June 1994 of 326 activist groups from 44 countries had called for an immediate moratorium on World Bank funded large dams until a comprehensive, independent review of all Bank funded projects had been conducted. International financial institutions were actually no longer able to fund further large dams in the face of public criticism. Enervated by steadily growing protests despite continual tightening of social and environmental standards, the institutions' representatives wearily likened the dispute, to a football match in which somebody kept moving the goalposts. But one proposal to emerge from the meeting in Gland was for all parties to work together in establishing the World Commission on Dams.

The WCD began its work in May 1998 under the chairmanship of Prof. Kader Asmal, then South Africa's Minister of Water Affairs and Forestry. Its 12 members were chosen to reflect regional diversity, expertise and stakeholder perspectives. But the Commission ran the risk of failure right from the start due to their confrontational attitudes. The members, who spent 2 years 6 months jointly organising hearings, consultations and case studies and analysing more than 100 existing large dams, could not be more disparate.

The Current Situation

A large dam is a dam with the height of 15 m or more from the foundation. If dams are 5-15 metres high and have a reservoir volume of more than three million cubic metres, they are also classified as large dams. Using this definition, there are more than 45,000 large dams around the world, almost half of them in China. They were built in the 20th century to meet the constantly growing demand for water and electricity. On a global scale, hydropower dams account for about 20 per cent of electricity generated, and in 24 countries, including Brazil, Democratic Republic of Congo, Zambia and Norway, hydropower covers more than 90 per cent of national electricity supply needs. Half the world's large dams were built solely or mainly for irrigation. Between 12 per cent and 16 per cent of world food production is based on dams, and as reservoirs they provide protection against floods.

Unfortunately, this impressive balance is counteracted by comparably significant problems. Construction of large dams is a major intervention in the ecosystem of rivers and the lives of many people. The WCD estimates that some 40-80 million people, mostly indigenous peoples, have been displaced by reservoirs worldwide and robbed of their livelihoods from fishing or farming. Serious conflicts are simmering between neighbouring countries because dams have turned off the water supply for downstream states.

The late Indian Prime Minister Jawaharlal Nehru once said: "Dams are India's new temples." Right up to the 1970s, large dams were seen as the synonym for development and economic progress. Dam-building reached its peak between 1970 and 1980, when an average of two to three new large dams per day were commissioned. But a considerable number of the dams analysed by the WCD have fallen short of their technical and economic objectives. Construction cost overruns averaged 56 per cent. Many dams have had negative ecological impacts, and the disadvantages for people living downstream were mostly not taken into account. The planning of dams did not examine sufficiently possible alternatives for meeting power and water needs. There were hardly any retrospective evaluations of dam projects.

A New Framework for Decision-making?

Despite this sobering stocktaking, the WCD arrives at an astonishingly simple finding: dams are primarily a means to an end. Their task is to improve the well-being of the people on a sustainable basis. This improvement should be economically acceptable, socially just and environmentally sound. If this goal can be achieved by a dam, its construction should be supported. Where alternative options offer a better solution they should be the preferred choice.

The Commission based its work on a set of five crore values for future decision-making: equity, efficiency, participatory decision-making, sustainability and accountability. With regard to legal aspects and the extent of the potential risks for those involved, the WCD proposes development of an approach based on recognising rights and assessing risks. All risk-bearers should have a place at the negotiating table.

The WCD also recommends seven strategic priorities for decision-making: gaining public acceptance; comprehensive options assessment; reviewing existing dams; sustaining rivers and livelihoods; recognising entitlements and sharing benefits; ensuring compliance; and sharing rivers for peace, development and security. These priorities are reinforced by 23 practical criteria and guidelines which can be adopted, adapted and applied by all actors involved in the dam controversy. For instance, the WCD suggests analysing points at issue together with the people affected by existing dams and developing joint proposals for solutions. People affected by new dam projects should be among their favoured beneficiaries, and their claims should be made legally binding.

The report offers a comprehensive compilation of knowledge, which previously was limited to individual case studies or a narrow specialist context, on the social, economic, technological and ecological problems and impacts of large dams. That makes the report a central reference which helps greatly in bringing objectivity into the debate. Provision of an analytical framework and strategic options is certainly an important step. But with regard to its task of developing internationally valid criteria and guidelines for the planning, design, appraisal, building operation, monitoring and shutdown of dams, the report remains very general. What will be decisive here will be to practice with the relevant actors the method and content of the suggested mediation process on the basis of specific cases.

Deeds Must Follow Words

The WCD's work must now be made useable for the private sector, civil society and development purpose. What is required is a discussion process involving all major actors. The objective of this process must be the development of practical and effective guidelines in addition to the current standards.

The WCD calls on bilateral development organisations and multilateral development banks to support only dam projects that have resulted from an open process of examining various options. The parties should observe the WCD guidelines. Measures to save water and power should be examined and, if applicable, be promoted.

Private sector companies should publish guidelines on corporate behaviour and acknowledge the WCD principles, criteria and guidelines. Further, the private sector should draw up and implement voluntary codes of conduct, management systems and certification procedures, such as the internationally recognised standard for environmental management (ISO 14001). The OECD's Anti-Corruption Agreement should be observed, and declarations of honesty incorporated in contracts. Business associations should develop processes to monitor compliance with the WCD guidelines.

NGOs should primarily check compliance with agreements and assist aggrieved parties to seek compensation. They should also assist in identifying relevant stakeholders for dam projects, using the rights and risks approach. Finally the NGOs should build up support networks and partnerships between them.

Importance of Information

But do these noble proposals provide the whole answer? The WCD process is based on the opportunities for personal development of every individual in an open society. But it is not enough to build on the negotiating abilities of the potentially affected alone because only specialists can anticipate the complex impacts of dams. Therefore participation presupposes that the mediation process contains a substantial informative component.

Relying solely on a mediation process is also not sufficient in providing for social impacts. There is no generally recognised method to determine the value of 'goods' in a subsistence economy. Without objective criteria, the moral demands on both sides are extremely high. Strategically-motivated behaviour will then not be prevented even if all participants agree readily that the subjective standard of living of people affected by a dam should be improved or at least maintained.

Mediation processes make sense only if agreements are observed. According to the WCD analyses, lack of compliance with agreements is the main cause of the negative social impacts of dams. In the case of dam projects co-financed by international donors, disbursements of funding installments could depend upon independent evaluations that confirmed compliance. Ensuring compliance is much more difficult in the case of projects financed by the private sector. Because there is no independent institution that could assume the role of arbitrator, it must be in the business world's own interests to act responsibly in ecological and social terms. To be credible, it must provide transparency and independent certifiers.

Due to the opposing interests involved, no-one should expect reaching consensus to be easy. But the example set by the WCD is not the only reason for hope. Taking a closer look at it, the model offers significant advantages for all participants. Partner countries and development organisations wish to continue to use the potential for development which dams will also offer in the future. The private sector will also continue to build and operate dams, and for that they need planning certainly. The advantages for NGOs and the people affected are also obvious.

4

End of Controversy on Large Dams

Was it worth the effort? The energy, the time, the money invested? Quite a number of people probably put this question to themselves on the 16th of November 2000 when Nelson Mandela launched the final report of the World Commission on Dams (WCD) in London. In an extraordinary process that lasted two and a half years, dam proponents and opponents worked intensely together. Twelve commissioners had tried hard to come up with a consensus on the effects of large dams in the past, and with recommendations for future sustainable planning on water and energy issues. And surprisingly enough: the Commission succeeded. The cost of ten million dollars was financed by governments, international agencies, the private sector, NGOs and various foundations—this is also a novelty. Dam-affected people, non-governmental organisations, companies, consultants, politicians, etc. contributed to the processes with their know-how and by giving support and additional resources to the Commission's work.

The establishment of the multi-stakeholder commission was the result of a growing and aggravating controversy about the social, ecological and, often enough, also economic costs of large dams.

Global Review Shows Faulty Planning Processes

The WCD Global Review of large dams proves that there were good reasons for massive resistance against large dams in the past. As the report says: "In too many cases an unacceptable price has been paid... especially in social and environmental terms, by people displaced, by communities downstream, by taxpayers and by the natural environment". While founders of large dams, political institutions, consultants and companies claimed in recent years that they had learned from their faults and improved their performance, the WCD highlights that, while policies and assessment procedures have been improved, it appears that business-as-usual too often continued to prevail:

- Even in the 1990s, impacts on downstream livelihoods were not adequately assessed or accounted for in the planning and design of large dams.
- Participation and transparency in planning processes for large dams was neither inclusive nor open, and while actual change in practice remains slow, even in the 1990s, there is increasing recognition of the importance of inclusive processes.
- Where opportunities for the participation of affected people and the undertaking of environmental and social impact assessment have been provided they often occur late in the process, are limited in scope and even in the 1990s their influence in project selection remains marginal.

Dams Hindered Human Development

The WCD states that "dams have made an important and significant contribution to human development, and the benefits derived from them have been considerable". A viewpoint that dam-affected people and NGOs hardly share, having in mind the experiences of the past. Medha Patkar, WCD Commissioner and activist in the struggle to save the Narmada river in India wrote in her comment to the Report: "Within the value framework the Commission propagates—equity, sustainability, transparency, accountability, participatory decision-making and efficiency—large dams have not helped attain, but rather hindered 'human development'.

The WCD also puts an end to the old viewpoint that the violation of human rights and the social costs that some have to pay can be justified by the benefits for the others. The idea that society as a whole would profit from so-called development that were benefiting from 'trickle down effect'. In fact this development model tends to aggravate social inequities and encourage environmental destruction leaving the rich better off, but the poor more marginalised and resentful. It is not a sustainable model.

It's Not Just About Dams

Although it is called the World Commission on Dams the issues before the Commission are much broader. The discussion about large dams is inevitably linked with the need to find solutions for supplying water and energy in the future. To do this in a sustainable way is one of the challenges of our time.

The WCD defined 5 core values that are based on internationally accepted norms like the Universal Declaration of Human Right to Development and the Rio Declaration on Environment and Development:

- Equity
- Efficiency
- Participatory Decision-making and Accountability

These values are not really new and some of them have been agreed upon decades ago, but the Global Review of the WCD showed that 'in real life' we are still far from implementing them.

The rights and risks approach of the WCD must be mentioned as an important tool: while funders and dam-builders talk a lot about their (mostly) financial risks, risks of the dam-affected people were not a big issue in the past. The Commission makes an important distinction: while the former take a voluntary risk and have the possibility to decide on whether or not they want to take it, the latter take involuntary risks and so far had hardly any option in deciding on whether or not they are willing to incur them. Rights must not be violated in order to serve the needs of other people. Respecting human rights is the minimum basis for discussion and is not negotiable.

To get better results in the future the WCD defined seven strategic priorities for decision-making:

- Gaining Public Acceptance
- Comprehensive Options Assessment
- Addressing Existing Dams
- Sustaining Rivers and Livelihoods
- Recognising Entitlements and Sharing Benefits
- Ensuring Compliance
- Sharing Rivers for Peace, Development and Security

According to the WCD there are several fundamental strategic points in the decision-making process. One point is right at the beginning of the planning stage: if the discussion about sustainable energy and resources management is carried out in a transparent, open and participatory process, further steps are much more likely to be accepted by all parties and will help prevent conflicts at later project stages. According to the WCD there are several fundamental strategic points in the decision making process. One point is right at the beginning of the planning stage: if the discussion about sustainable energy and resources management is carried out in a transparent, open and participatory process, further steps are much more likely to be accepted by all parties and will help prevent conflicts at later project stages.

The strategic priority 'Addressing existing dams' includes (among other policy principles) the identification and assessment of outstanding social issues associated with existing large dams. Considering that between 40 and 80 million people have been displaced by large dams (a lot more have been directly or indirectly affected) this is a difficult task. Financial institutions, development agencies and companies are a lot more interested in looking at the future than dealing with the complex problems of the past. But there is no way out. Dam-affected people of the past along with future dam-affected people are unlikely to accept new dams and believe in what is being promised while the legacy of the past is being forgotten. To solve these problems is a condition for further constructive discussions if not a mere moral obligation. How can those responsible for planning future dams think of gaining public acceptance if not by showing that they mean what they say?

The recommendations of the WCD are of fundamental importance, because they were agreed upon in a multi-stakeholder process. In other words, they were made by representative of industry as well as those from dam-affected people, by members coming from industrialised countries as well as those from developing countries. Within the Commission, it was possible to integrate the most diverse views and to come to a consensus.

No More Dams?

Although in their assessment the evidence would have allowed even stronger recommendations, NGOs and people's movements welcomed the report and asked for immediate implementation. Other stakeholders do not seem to be very enthusiastic about it. Quite a few representatives from industry and investors seem to fear that the implementation of the WCD recommendations would lead to an abrupt stop in dam building.

All appreciate that the WCD report vindicates many concerns raised by NGO campaigns. Given the role of financial institutions in funding large dams and in the WCD process, and based on the WCD report's recommendations, all have to call on all public financial institutions, including the World Bank, the regional development banks, the export credit agencies and bilateral aid agencies, to take the following actions:

- All public financial institutions should immediately and comprehensively adopt the recommendations of the World Commission on Dams, and should integrate them into their relevant policies, in particular those on water and energy development, environmental impact assessment, resettlement, and public participation. In particular, as recommended by the WCD, no project should proceed without the free, prior and informed consent of indigenous people, and without the demonstrable acceptance of all those who would be affected by the project.
- All public financial institutions should immediately establish independent, transparent and participatory reviews of all their planned and ongoing dam projects.

While such reviews are taking place, project preparation and construction should be halted. Such reviews should establish whether the respective dams comply, as a minimum, with the recommendations of the WCD. If they do not, projects should be modified accordingly or stopped altogether.

- All institutions which share in the responsibility for the unresolved negative impacts of dams should immediately initiate a process to establish and fund mechanisms to provide reparations to affected communities that have suffered social, cultural and economic harm as a result of dam projects.
- All public financial institutions should place a moratorium on funding the planning or construction of new dams until they can demonstrate that they have complied with the above measures.

Irrigation Management:

Facing the Challenge

Irrigated agriculture is up against an enormous challenge. India's population continues to grow at a tremendous rate. Over the next 10 years, there will be more extra mouths to feed. Water is becoming increasingly scarce for agriculture, with conflicting demands being made on limited supplies by the domestic and industrial sectors. The most attractive irrigation sites have already been exploited. Yet, irrigated agriculture will have to deliver average output increases of at least 3.5 per cent per year if future food demands are to be met in India.

Forty years ago in the 1950s and 1960s, India has worried about its capacity to produce sufficient food to feed its growing population. Episodes of food scarcity were not uncommon. It was dreaded that millions would die of hunger. Then dawned the miracle of the "green revolution". Cereal production increased by leaps and bounds, boosted by expanded irrigation, increased fertilizer application and modern crop varieties. The vigorous response in mobilising financing towards boosting agricultural production paid off.

Given the importance of irrigation to the India's food supply and the vast resources already expanded on irrigation development, it is tragic that the actual performance of irrigation systems is so disappointingly low. In the post-green revolution era, it has become increasingly evident that the performance of irrigation systems, especially large-scale systems, is suboptimal whether measured in terms of achieving planned area targets or in terms of production potentials created by the physical works.

In many irrigation systems, the actual area irrigated is much less than the command area. Sharp inequities in water supplies between farmers in the head reaches of irrigation systems and those located downstream is another manifestation of poor performance. Investigations in the Tungabhadra Irrigation Scheme reveal that the tail-end of a major distributary commanding 25 per cent of the total area, received approximately 20-40 per cent of the targeted discharge while the upper reaches got more than their share.

Lack of maintenance has caused many systems to fall into disrepair, further inhibiting performance. Over time, distribution canals have become silted up, increasing the likelihood of breaching, damage to outlets and leading to salt build-up in the soil.

Successful Farmer-managed Irrigation Systems

Farmers have long demonstrated their potential capability to manage irrigation systems efficiently. Farmer-managed irrigation systems (FMIS), also known as traditional, indigenous, communal or peoples' systems, are often classified "minor" or "small-scale" irrigation systems, although they may be found in command areas of 15,000-20,000 hectares.

Many successful farmer-managed irrigation systems which have been functioning effectively for hundreds of years, represent a valuable, accumulated investment. They are also reservoirs of largely untapped irrigation management experience.

Research has revealed that FMIS contribute to the production of a significant portion of the subsistence food supply. Ground-water irrigation systems, often found in areas affected by drought, play a strategic role in promoting food security. In India ground-water development which is increasing in importance is predominantly farmer-managed. In this country an estimated 20 million hectares are already under ground-water irrigation. When mismanaged though, ground-water irrigation could impact negatively on the environment.

In addition to the hectarage under farmer-managed ground-water irrigation, farmer-managed tank irrigation systems cover about 8.5 million hectares in this country. Farmer-managed irrigation systems have also allowed intensification of agriculture to partially meet the food needs of rapidly growing populations.

Little Awareness

Yet, despite the widespread interest in irrigation management turnover throughout India, there is very little documentation about the processes used and the results obtained from irrigation management turnover. Many policy-makers do not know how to turn over management of their irrigation institutions in an effective way. They typically have very little, if any, awareness of the range of organisational options which may be suitable under different conditions. There is an urgent need, therefore, for a systematic, comparative assessment of the range of approaches being used, constraints to implementation and the impacts on performance of transferring management to non-governmental or farmers' organisations.

Successful irrigation in the future will be that which supports much higher levels of agricultural productivity, enhances responsiveness to more diversified and dynamic crop markets, stimulates more profitable irrigated agriculture for wide numbers of rural poor, substantially improves water use efficiency and supports and sustainable use of scarce land, biomass and water resources.

In the coming years the irrigation sector will be in ferment, with decision-makers, agency managers and farmers needing better information and strategic processes to make intelligent choices in the management of irrigated agriculture.

6

Strategies for Improved Water Management

Since all surface and sub-surface water related activities are closely interlinked through upstream-downstream relations in river-basins and aquifers, the whole river-basin must be considered in all water development policies and research.

River-basin authorities should be established, for both national and international basins, and should have regulatory powers over inter-sectoral allocation of water, enforcement of water quality standards, arbitration in disputes, and compensation procedures. International basin authorities should be composed partly of representatives of national basin authorities.

Water policies should encompass underground water, surface water, and direct utilisation of rainfall, as an integrated continuum requiring common strategies.

Participation of Users and Stakeholders

Governments should strengthen stakeholder participation in the allocation and management of water resources by means of:

1. Identifying and establishing water rights for all users, following economic, cultural, environmental and social considerations;

2. establishing compensation procedure (both national and international);

3. using impact assessment as a tool to identify livelihood rights and ecological considerations.

Stake-holder participation in water allocation and management should be used to reconcile multiple or contradictory objectives and to reduce conflicts over ownership or control of water.

Environmental Protection

To guide development actions and to monitor changes, basin-wide mapping of "hot spots" (areas of special problems or vulnerability with respect to water quality) should be undertaken; sources of major pollutants should be identified and published. Water bodies which have special ecological or societal significance should be preserved.

Training of professionals for the water sector should develop new environmental dimensions, and should address such issues as; conservation of water; responses to water scarcity; disaster preparedness.

Research into the Impacts of Water Policy

The impact of water policies, and policy changes, are not well understood. There is a need for structured research on impacts rather than reliance on anecdotal evidence.

Water development actions can have negative effects on equity, and can injure individuals or groups even while bringing wider benefits. There is a need for analysis of responses and solutions to these effects.

Externally defined systems of water rights will lack legitimacy and will fail to meet the needs of all users, if they do not take account of users' perceptions and customs. National research institutes should therefore develop and apply methodologies for assessing users perceptions and customary rights.

Capacity-building for Integrated Management

The development of river-basin authorities will require the training of their personnel in: coordination; planning of systems for monitoring and evaluation; sensitisation to new responsibilities; integrated land and water resources uses; use of modelling tools; implementation of guidelines; procedures for stake-holder participation.

Politicians and policy makers need to become more aware of issues concerning; water scarcity and water needs; equity of access to water; water rights; sustainability; service improvement; enhancing the performance of water systems; and stakeholder participation.

External Assistance for Development

Partnerships or "twinning" arrangements, should be promoted between new riverbasin authorities and existing successful ones.

External donor assistance programmes should focus on users' groups, especially among the disadvantaged and the poor. They should take a basin perspective, recognising existing institutional arrangements and inter-sectoral requirements. Within this frame work they should foster and encourage mechanisms for devolution of responsibilities, tasks, and authority to local level users' groups.

External help is needed to develop and evaluate water data-bases, agricultural knowledge systems, methodologies and processes for water development and evaluation, and other information areas.

Emergencies and Conflict Reduction

National governments and river-basin authorities should develop contingency plans for the management of their responses to natural disasters, prolonged droughts, and other emergency situations in which normal procedures of water allocation and management may require temporary modifications.

Policies and actions in regard to shared water sources should be improved, with emphasis on consultation, coordination and sharing of information.

7

Water Problem in South India

Southern India's fast-growing urban areas and its farmers will collide over water allocation unless the government and water users take swift action. Urgently needed are measures that ensure efficient water use in cities and on farms, including regulation of groundwater withdrawal, restoration of traditional rain-collecting reservoirs, and experimentation with different cropping strategies.

In India's arid southern tip, typifies the plight of the country's extensive drylands. Scanty rainfall means residents often must get by on water drawn from small reservoirs and underground sources.

Extended dry periods are punctuated by intense monsoon bursts. When rains lash South India during the monsoons, only a fraction of the downpour is captured for later use. The rivers are seasonal and small compared to the Himalayan cataracts up north.

During the monsoon, these waters swell briefly to gigantic proportions recharging the water tables in their basins and then subside. In times of drought, when the rivers are narrow ribbons, water drawn from below ground sustains crops.

In the cities of South India the problems of water supply are exacerbated by antiquated or nonexistent infrastructures that cannot keep up frenetic growth.

Farmers in South India depend on irrigation to see them through the growing season. They draw their water from three sources: Government canals that bring water from rivers and dams, traditional reservoirs known as "Tanks", and public and private wells often fitted with electric pumps. All three need to be made more efficient.

Besides being enormously expensive, large surface irrigation canals and dams are plagued by massive leaking and evaporation. Tanks collect rain water runoff behind small earthen dams. Bu these are falling into disrepair as farmers take advantage of low interest government loans and heavily subsidised electricity to switch to private wells fitted with electric pumps. Without maintenance, the tanks fill up with silt and hold less water.

As more wells are dug and as percolation tanks that once recharged underground water supplies fall into disuse, the water table is dropping at a rapid rate. Farmers who can afford to install powerful electric or diesel pumps on their wells are able to tap the retreating water table, but poorer farmers who raise their water by hand or cattle power from shallow wells often come up dry.

No regulations govern the amount of water a farmer can withdraw from a well, so underground reserves are available on a first-come-first-served basis. Add to this situation is electricity rates based on the horsepower of pumps, not on the amount of electricity they consume, and there's no incentive to save water.

Any effort to limit groundwater extraction carries heavy political liabilities. Politicians are reluctant to risk raising the ire of farmers, who provide the majority of their votes.

Preventing overdraft where water tables are falling would be a first step toward correcting water troubles. While some areas face the prospect of rationing, others have an abundance of groundwater still to be tapped. Effective legislation must be based on detailed and accurate map of groundwater supplies—something that doesn't currently exist.

They should also consider instituting higher user fees for its public canal projects and returning to its old policy of charging farmers according to the electricity they consume. Making farmers pay closer to the full price for water and electricity would provide revenue for digging communal wells equipped with electric pumps and overhauling the tank system, while providing an incentive for farmers to conserve.

Using tanks and wells together maximises the effectiveness of both, since tanks take pressure off groundwater supplies and wells sustain crops during their final weeks of growth when tanks are low. Any efforts to limit the number of new wells dug has to include measures that provide something to those without water, otherwise only those who currently own wells will benefit from groundwater conservation.

Officials might also experiment with encouraging farmers to plant less water-demanding crops such as ragi, sorghum and pulses in place of cotton, sugarcane, bananas and spices.

If politicians act now, they still have a chance to craft measures that are fair rather than desperate. The time is nearing when water for drinking and irrigating crops will have to be culled from careful conservation rather than from the ground.

8

South Asia Quarrels Over Water

Bangladesh, Bhutan, India and Nepal have at least three features in common—they are all situated on the southern slopes of the Himalayas, they have high levels of poverty, and they are all rich in water resources. To these add a fourth: they frequently quarrel about water.

For decades, policy makers in the four countries have grappled with the problem of how to harness the rivers that flow from the mountains and serve the region's over one billion people. Mostly, they have gone in for large dams and grand irrigation schemes. But, though these promise a lot, their usefulness is being increasingly questioned by many independent engineers, scientists and social activists. And they are ought to contribute to water conflicts between countries.

Large hydro-power projects represent a development model that is biased toward industry and urban areas. And irrigation schemes favour rich farmers with large farmlands, most useful for water and fertilizer-intensive cash crops that poor peasants cannot afford to grow. Because of their size, they also cause huge displacements—again, of poor farmers. Social activists say that while such projects may have increased overall food production, they have also increased the rich-poor gap in a region that is home to the world's largest number of absolute poor-people living on incomes of less than one dollar a day.

In tandem, many South Asian hydro-power engineers and economists say 'national' and top-down engineering solutions to water management ought to be replaced by a new 'basin-wide' or regional approach through environmental and socially benign projects.

One example of such 'national solutions' is the Indian Farrakka Barrage on the Ganges. Built near the Bangladesh border in order to divert water to the Calcutta port, it has led to problems with falling water tables and salinity downstream in Bangladesh. And although India and Bangladesh resolved the long-simmering dispute in 1996, Dhaka is now proposing another expensive Ganges Barrage to solve the problems created by the first barrage.

Of the four countries, Nepal has the highest per capita potential for hydro-power generation—its narrow valleys and steep terrain mean water flows faster, which is good for power generation. But although Nepal could potentially turn its rivers into 'hydro dollars', Himalayan dams are expensive to build and the country's experience in dealing with India on joint river projects has been patchy. At the same time exporting power to India, where powercuts that can last up to 12 hours at a time in some cities, means foreign exchange for Nepal.

Because many Nepalese perceive past border irrigation projects as unfairly benefiting India, joint Indo-Nepal river projects are a political issue. A recent treaty to harness the Mahakali River on Nepal's western border with India is a case in point.

Although India and Nepal agree on equally sharing the 7,500 megawatts of power from the project, negotiations have been stalled over Nepal's demand that it be compensated for downstream irrigation benefits to India. Nepal is not allowed to harness water in many of its river for its own use without India's concurrence as these rivers flow into India, and Nepal is bound by past bilateral treaties.

As one official at Nepal's Water and Energy Commission put it: "What's in it for Nepal? Why should we write off benefits from the Mahakali for future generations of Nepalis? Things look even bleaker for larger joint projects like a planned 10,800 mw dam on the Karnali River—the $10 billion Chisapani dam project is expected to displace some 60,000 Nepali farmers. In addition,

India has already used up all the natural flow of the Karnali for its own irrigation downstream.

Things are different when it comes to Bhutan: India and the tiny Himalayan kingdom have sorted out their river projects without any major difficulty although experts say it is too early to assess how much the Bhutanese will really benefit. According to an estimate by an Indian water consultant, Bhutan can generate up to 20,000 mw of power from its rivers.

Generating funds for large projects does not seem to be a problem. Traditionally, the World Bank has been one of the major financiers for dams worldwide having undertaken 400 projects involving dams since 1970. In 1988, India alone had about 473 dams under construction, with the Bank involved in 45 of them.

When the World Bank recently pulled out of some controversial mega projects, such as the Arun III dam in Nepal, multinational companies stepped in. In Nepal, three medium-sized hydropower projects worth more than 300 million dollars are being built with foreign direct investment.

FDIs are backed by organisations like the Japanese-managed Global Infrastructure Fund (GIF) which is interested in putting money into such projects as the 270 metre-high Kosi High Dam in Nepal, the Ganges Barrage in Bangladesh and the mammoth 20,000 mw Dhihang dam in India. Funds also come from the Manila-based Asian Development Bank.

But many resource economists maintain that the chances of making colossal mistakes is higher with large projects. It is all a question of risk-management. Can our countries afford to take the risk of spending billions on doubtful projects that can go dangerously wrong?

9

Solutions for A Water-Short World

As populations grow and water use per person rises, demand for fresh water is soaring. Yet the supply of fresh water is finite and threatened by pollution. To avoid a crisis, many countries must conserve water, pollute less, manage supply and demand, and slow population growth.

Caught between growing demand for fresh water on one hand and limited and increasingly polluted water supplies on the other, many developing countries face difficult choices. Populations continue to grow rapidly. Yet there is no more water on earth now than there was 2,000 years ago, when the population was less than 3 per cent of its current size. Rising demand for water for irrigated agriculture, domestic (municipal) consumption, and industry are forcing stiff competition over the allocation of scarce water resources among both areas and types of use.

Today 31 countries, accounting for under 8 per cent of the world population, face chronic freshwater shortages. By the year 2025, however, 48 countries are expected to face shortages, affecting more than 2.8 billion people – 35 per cent of the world's projected population. Among countries likely to run short of water in the next 25 years are Ethiopia, India, Kenya, Nigeria, and Peru. Parts of other large countries, such as China, already face chronic water problems.

In much of the world polluted water, improper waste disposal, and poor water management cause serious public health problems. Such water-related diseases as malaria, cholera, typhoid and schistosomiasis harm or kill millions of people every year. Overuse and pollution of water supplies also are taking a heavy toll on the natural environment and pose increasing risks for many species of life.

What Can Be Done?

It may already be too late for some water-short countries with rapid population growth to avoid a crisis. Many other countries can avoid the coming crisis if appropriate policies and strategies are formulated and acted on soon. Whether water is used for agriculture, industry, or municipalities, there is much room for conservation and better management. Effective strategies must consider not only managing the water supply better but also managing demand better.

To avoid catastrophe over the long-term, it also is important to act now to slow the growth in demand for fresh water by slowing population growth. Currently, in many developing countries millions of people want to plan their families and to use contraception. Family planning programmes have played an important role in assuring individual reproductive health and in reducing national fertility levels. Continuing and expanding these programmes also can help assure that population growth eventually slows to sustainable levels in relation to the supply of freshwater.

Towards a Blue Revolution

The world needs a Blue Revolution to conserve and manage freshwater supplies in the face of growing demand from population growth, irrigated agriculture, industries, and cities—just as the Green Revolution transformed agriculture in the 1960s. A Blue Revolution will require coordinated responses to problems at local, national, and international levels.

Locally led initiatives show that water can be used much more efficiently. When communities manage freshwater resources efficiently, they also manage other natural resources better, improve sanitation, and reduce disease. At the national level, especially in water-short regions with dense populations, adopting a watershed or river-basin management perspective is a needed alternative to uncoordinated water-management policies by separate jurisdictions. At the international level countries that share river basins can fashion workable policies to manage water resources more equitably. Development agencies need to focus more on assuring the supply and management of freshwater resources and on providing sanitation as part of development and public health programmes.

A water-short world is an inherently unstable world. As the next century dawns, water crises in more and more countries will present obstacles to better living standards and better health and even bring risks or outright conflict over access to scarce freshwater supplies. Finding solutions should become a high priority now.

Solving Conflicts over Water Uses

There are few issues that have greater impact on the life of mankind and the planet as a whole than the management of our most important natural resource water. This has only been realised in detail more recently by the general public as well as by many planners and decision-makers. Around the world they have begun to appreciate the critical importance of a reliable water supply for their future survival and sustainable development. Rivers are lifelines in countries like India serving different uses such as transport, agriculture, fisheries, personal hygiene and others. Conflicts over use of water resources must be settled through better water management policies.

In many localities of the Earth water-related problems have become extremely acute, even critical. In some places they are the source of social instability and are a threat to international security. There is not the slightest doubt that, with further population increase and under the "water business-as-usual" scenario, these problems will become ever major acute, thus creating ever more instability. After decades of water waste, water pollution, and inability to provide basic water services to the poor we must fundamentally change the way we think about and manage water. We have to realise that water can no longer be considered to be a cheap and plentiful resource, which can be used, abused or squandered without much concern for further human welfare.

As fresh water is becoming scarce, that is when there is not enough water to satisfy all demands, competition develops. Besides the well known tensions at international level over limited water supplies there is an increasing and in many cases a far more important competition over water arising within countries, between sectors. Such competition, for instance, occurs between farmers for irrigation water and between farmers and non-agricultural users of water such as cities (including industries and power plants) and environmental concerns (recreation, fish and other wildlife). Because of this increasing competition irrigated agriculture around the world, but especially in developing countries, faces important challenges in the coming decades. On the one hand, it has to provide a major share of the required increases in food and fiber production to meet the objectives of poverty alleviation and development, on the other it is threatened by water shortages arising out of increasing competition from domestic, industrial and other sectors. This situation is further worsened by dwindling financial resources available for capital expenditures as the cost of new irrigation schemes increase.

Land under Irrigation

On a regional basis, it is estimated that, for example, around 60 per cent of the value of crop production in Asia is grown on irrigated land. The irrigated sector performs an essential task in meeting the basic food needs of billions of people. It has provided more than half of the two most important basic staples and close to a third of all food crops. In the future, the irrigated sector will have to provide an even larger proportion of the total food output. The question arises whether irrigated agriculture will be in the position to provide the extra food needed to feed a growing population despite an increasing water scarcity and inter-sectoral competition. There is no general, no easy answer to this question. But the following implications are foreseeable.

The irrigation sector has to recognise that economic structures are dynamic and not static. The economics of many countries have undergone considerable change in the last decades. In connection with this change agriculture is losing its leading role in economic development and this is, besides other things, affecting the allocation of water resources between agriculture and other users. The same applies to many other countries around the world, especially developing countries in arid and semi-arid climates.

New Water Policies are Needed

Because of the growing competition for ever-scarcer water resources, governments and water authorities are forced to change water policies. Objectives of the new policies are demand-decreasing and demand-shifting. Irrigated agriculture, by far the dominant water user, will be strongly effected by such policy changes.

Water has an economic value in all its competing uses and should be recognised as an economic good. Agriculture will in future not any more get water free of charge as it did in the past. Farmers will have to pay for the water, and costs will be increasing steadily over the years to come.

There will be a reduction of the role of governments in rural water projects and an increasing importance of local user groups. Experience shows that an important solution to water related problems is to give users the responsibility for developing and managing the water resource. This requires that farmers become properly organised in water user associations, able to discharge their new responsibilities, and that they have security of tenure to the land they farm and irrigate.

During the recent decade, growth in crop productivity in irrigated areas has slowed and competition for water for non-agricultural uses has increased. These developments place strong demands to develop water resource policies to maintain growth in irrigated agricultural production:

- facilitate efficient allocation of water across sectors and final demands; and
- reverse the ongoing degradation of the water resource base, including the watershed irrigated land base, and water quality.

What water policies can lead to efficient increases in irrigated production while reducing resource degradation in the irrigated areas in developing countries and releasing water for growing non-agricultural demands? What policies can be implemented to conserve water in non-agricultural uses to reduce competition between sectors?

Questions to be Answered

Among others the following questions still have to be answered.

1. What are the implications of growing competition between agricultural and non-agricultural uses of water for the availability and productivity of water in agriculture?

2. How to use regulations, water prices, pollution taxes and effluent charges to encourage water conservation and pollution control in industries and households?

3. What are the production, equity, employment generation and income impacts of alternative water allocation mechanisms in different agroeconomic and scarcity environments? What investment and administrative costs are associated with different mechanisms? What is the impact of resource allocation methods on water use, cropping patterns, crop yields, and fertilizer and other input use, capital investments, farm income, and environmental degradation?

4. What is the connection between alternative allocative mechanisms and the environmental externalities caused by irrigation, including water logging, salinisation, ground water recharge and ground water mining?

5. How will food security in low income countries be effected by changing water policies?

In the context of the growing competition for water it is important to enable policy makers to evolve a policy-mix which will result in efficient, equitable and environmentally sustainable allocation of water resources to user sectors. To achieve this, extensive research work is needed.

11

The Coming Water Crisis

Freshwater is emerging as one of the most critical natural resource issues facing humanity. The world's population is expanding rapidly. Yet there is no more freshwater on earth now than there was 2,000 years ago, when the population was less than 3 per cent of its current size.

Water is, literally, the source of life on earth. The human body is 70 per cent water. People begin to feel thirst after a loss of only 1 per cent of body fluids and risk death if fluid loss nears 10 per cent. Human beings can survive for only a few days without freshwater. Yet, in a growing number of places people are withdrawing water from rivers, lakes, and underground sources faster than they can be recharged—"unsustainably mining what was once a renewable resource," as one researcher puts it. Currently, 31 countries—mostly in Africa and the Near East—face water stress or water scarcity.

Population growth alone will push an estimated 17 more countries, with a projected population of 2.1 billion, into these water-short categories within the next 30 years. By the year 2025, 48 countries, with more than 2.8 billion people—35 per cent of the projected global population in 2025—will be affected by water stress or scarcity. Another nine countries, including China and Pakistan, will be approaching water stress.

Beyond the impact of population growth itself, the demand for freshwater has been rising in response to industrial development, increased reliance on irrigated agriculture, massive urbanisation, and rising living standards. In this century, while world population has tripled, water withdrawals have increased by over six times. Since 1940 annual global water withdrawals have increased by an average of 2.5 per cent to 3 per cent a year compared with annual population growth of 1.5 per cent to 2 per cent. In developing countries over the past decade water withdrawals have been increasing by 4 per cent to 8 per cent a year.

Moreover, the supply of freshwater available to humanity is shrinking, in effect, because many freshwater resources have become increasingly polluted. In some countries lakes and rivers have become receptacles for a vile assortment of wastes, including untreated or partially treated municipal sewage, toxic industrial effluents, and harmful chemicals leached into surface and ground waters from agricultural activities.

Caught between finite and increasingly polluted water supplies on one hand and rapidly rising demand from population growth and development on the other, many developing countries face uneasy choices. The lack of freshwater is likely to be one of the major factors limiting economic development in the decades to come.

Slowing Demand, Conserving Supplies

To avoid a water crisis, particularly in water-short countries with rapid population growth, it is vital to slow the growth in demand for water by managing the resource better, while at the same time slowing population growth as soon as possible. Family planning programmes play an important role not only for individual reproductive health but also for sustainability of the use of freshwater and other natural resources in relation to population size.

As population grows, so does demand for freshwater for food production, household (municipal) consumption, and industrial uses. The availability of freshwater limits the number of people that an area can support and affects standards of living. In turn, population growth and density typically affect the availability and quality of water resources in an area, as people attempt to assure their water supply by digging wells, constructing reservoirs and dams, and diverting the flow of rivers. If needs consistently outpace available supplies, at some point overuse of water leads to the depletion of surface and groundwater resources, triggering chronic water shortages.

Scarce and unclean water supplies are critical public health problems in much of the world. Polluted water, water shortages, and unsanitary living conditions kill over 12 million people a year.

Competition for freshwater supplies breeds social and political tensions. River basins and other water bodies do not respect national borders. For example, one country's use of upstream water often subtracts from the supply available for use by downstream countries. As the 21st century dawns, there is a growing risk that wars will be fought over access to freshwater supplies.

If a crisis is to be averted, the world's overuse and misuse of freshwater must end as soon as possible. We cannot afford to keep wasting and fouling our precious supplies of freshwater. Increasingly, human activities are altering the flow of water and drawing down freshwater supplied faster than they can be replenished. Throughout the world enormous amounts of water are wasted due to inappropriate agricultural subsidies, inefficient irrigation systems, leaky municipal pipes, improper pricing of municipal water, poor watershed management, and other imprudent practices. It is time for widespread conservation measures, effective water management policies, and growing attention to assuring freshwater supplies and decent sanitation as part of development and public health projects.

Water:

Will There be Enough?

Humankind has a special relationship with water. In every civilisation, the most ancient traditions associate this precious resource with the origins of life, purification and regeneration. Far form being a mere raw material such as oil, water is vital for life, indispensable to the economy and so rich in symbolic value that it triggers passionate responses. All the computers in the world will never be able to express the real perception of the value of water or codify the interaction between it and people.

For decades, experts have been making grim forecasts that the Earth will start running out of water and that conflicts over this precious liquid will erupt into wars. The situation is indeed alarming.

How much water is there in the earth's reserves? Highly expensive probes have been sent to the Moon, Mars and the satellites of Jupiter and Saturn to find out whether there is water on them, but we still lack accurate data about the earth's hydrological resources. Such information would help to provide a cleaner picture of the future and, especially to foresee the global repercussions of demographic and climate change.

One thing we know about water is that there is plenty of it. The total volume is put 1.4 billion cubic Kilometres—which could be imagined as a 2,650-metre-deep layer of liquid evenly disturbed over the entire surface of the planet. But 98 per cent of it is salt water, mainly in the oceans and seas. Most of the earth's freshwater is trapped in the polar ice caps. Less than 1 per cent of it is available in lakes, rivers and shallow easily-accessible aquifers. These water resources are constantly in flux. Water from the oceans and land evaporates into the atmosphere before falling again as rain or snow, nourishing plants and swelling rivers that flow into the sea. It also seeps through the ground and percolates down to acquifers. Very deep groundwater, known as fossil water, is impervious to seepage and not renewable.

In the industrialised countries, all you have to do is turn a tap and before you know where you are you've used a considerable amount of water up to 600 litres per person a day in the United States. In hot developing countries, where shanty-towns on the edges of cities are crowded with growing numbers of migrants from the rural areas, a spigot and two litres of water a day are a luxury.

Over 1.3 billion people received improved drinking water services and some 750 million got better sanitation facilities during the International Drinking Water Supply and Sanitation Decade (1980-1990). Approximately 1.2 billion people still have no access to drinking water and 2.9 billion lack sanitation. The resulting water-borne diseases take the lives of five million people a year, most of them children.

Farming and manufacturing account for most of the world's water consumption, far outdistancing human needs of the 3,240 Km^3 of fresh water drawn every year, only 8 per cent are used for human consumption. Each year fewer than ten countries use 60 per cent of the world's 40,000 billion m^3 of surface and ground water. Lastly, per capita consumption rises with the standard of living, ranging from 260 litres a day per person in Israel to 200 in Europe, 70 for a Palestinian on the West Bank and 30 in Africa.

The Dangers of Irrigation

Demand runs highest in places where irrigation is indispensable, such as central Asia, Iraq, Iran, Pakistan Madagascar and also in some industrially developed countries such as the United States. Farming accounts for two-thirds of the total water resources used by humans—a figure that rises to 80 per cent in the Southern countries. Developing countries consume twice as much water per hectare to irrigated land as industrialised nations, yet their production is three times lower.

Because of the heat, half the water evaporates in storage areas or when flowing through open-air irrigation canals. Poorly conceived irrigation projects lead to deterioration of the soil.

The first is Pakistan, during the first half of the twentieth century, 10 million hectares were abundantly irrigated in the Indus plain. Water logging caused by irrigation, combined with a high rate of evaporation, has led to salinisation of the soil, making it unproductive. The second example, the Aral Sea in the former Soviet Union, is different but the result is the same. Much of the water from the Syrdarya and Amu Darya rivers that flow into the huge lake has been diverted to feed 1,80,000 Kms. of irrigation canals, only 12 per cent of which have been made watertight. The rivers' flow is considerably restricted and the Aral Sea is drying up. Irrigation for agricultural purposes is expensive. To make it profitable farmers must use massive amounts of pesticides, herbicides and fertilizers on increasingly exhausted soil. The impact of pesticides on health has been-over looked. The child morbidity and mortality rates are among the world's highest.

Water:

An Educational and Informative Approach

The most characteristic element of our planet is undoubtedly water. Indeed, more than two-thirds of the earth's surface is covered by water—the total volume representing almost 1,500 million cubic kilometres. About 94 per cent of this water is found in the oceans, almost 6 per cent is located underground and in glaciers, whereas rivers, lakes, soil moisture and atmospheric vapour, which constitute the major source of drinking water, account for a mere 0.0221 per cent of the total volume.

Water is indispensable for all living organisms. Life, as we know it, is impossible without water. It is present in all aspects of our life, directly or indirectly, next to the air we breathe, and together with the soil that we live upon, water constitutes the most important part of our environment, our most precious resource. And yet, except in the arid or semi-arid regions of the world, its value is generally overlooked until some catastrophe—natural or man-induced—force our attention to its worth. But even so, no sooner is the situation remedied than, more often than not, we revert to our old attitude.

The reason for this sort of indifference is undoubtedly attributable to the fact that, except in exceptional circumstances, water has always been considered as a "gift of the gods", as something that human beings are as naturally entitled to as the air they breathe. Its supply, however uneven, has always seemed inexhaustible because water has a natural regenerative cycle which, until the present century, was beyond human control or interference—or even proper comprehension. But the trend of social, political and economic evolution, notably in the past 200 years, with an increase of industry, agriculture, technology, and above all, a vertiginous population growth, as led to a dramatic revision of the age-old belief that no demands made by human populations on the natural resources of the planet are in the process of setting in motion vicious circles in the environment from which it is becoming increasingly difficult to extricate ourselves, not only as concerns the present, but far more important, for the future. Thus, the overuse—or abuse—of water resources has started affecting seriously not only the water cycle but the very nature of water in such a way that, in conjunction with other abuses of the environment, the results have been climate changes, droughts, flooding, desertification on the one hand and acid rain, water pollution and eutrophication on the other.

Actually, the problem of water is to be considered less in terms of quantity—though with a steeply increasing world population making increasingly heavier demands on a fixed quantity of water, one will sooner or later be confronted with this aspect of the problem too—than in terms of proper distribution of available resources taking into account sound management, stock-age and maintenance of quality. For among the major preoccupations of humanity in the coming years, adequate supply of freshwater to the teeming populations figures in the forefront. Between 1900 and 2000 water consumption will have globally increased tenfold and though the share of agriculture, the major consumer of freshwater, is expected to drop significantly (from 90 per cent to 62 per cent approximately), that of industry and the cities will have increased enormously (approximately, from 6 per cent to 24 per cent and 3 per cent to 8 per cent respectively).

Given the current trend of societal evolution, i.e. greater emphasis on industry and increasing migration towards the cities added to the global population boom, these figures are certainly cause for concern. Not only because of the damages caused to freshwater resources through the increasing use of fertilizers in the search to maximize agricultural production to cater to the increasing populations, but equally because the mushrooming of industries and urban concentrations are sources of increasing water pollution. Though the industrialised nations have more than their fair share of blame in this matter insofar as the current state of water pollution goes, for the future, it is in the developing world that lies the major source of concern. Lack of resources for adequate urban planning, the increasing role of industry in the search for economic solutions added to uncontrollable population pressures are already on the way to creating an explosive situation in a great number of developing nations with the available water supply becoming more and more inadequate in terms of quantity as well as quality. And then one considers the fact that around 80 per cent of all diseases are estimated to be water related, and that by the year 2000, 51 per cent of the world population will be urban based, one can hardly be accused of exaggeration in speaking of an explosive situation.

Attacking such a vast problem is no mean task. Water being the very source of life, what concerns water concerns every aspect of life. Thus, be it climate change, pollution, desertification, deforestation, food production....or whatever other major environmental problem that humanity is confronted with today, water constitutes one of the prime factors. Managing our water resources with care and intelligence for the use of present and future generations is a major responsibility which has to be shared by governments and the public alike, for no sector alone can deal efficiently with so vital a problem which affects not only the present but also the future of humanity. Again, as in the case of biodiversity and climate change, the problem of water being a global problem, international cooperation is of utmost importance since activities in one part of the planet are likely to produce consequences in other regions of the world. Concerted action by the international community alone is capable of dealing effectively with a problem of such far-reaching consequences

If our planet is to be saved from disaster—for in jeopardising our water resources we are guilty of nothing less than condemning life itself on our planet—we have to work for sustainable results: short-term plans for the present which will dovetail into medium-term ones for the coming generations without compromising the possibility, at the same time, of careful long-term planning to guarantee the future of the planet. In this, the part of environmental education and information of the people is fundamental. No strategy, no policy, no plan—be it ever so well prepared and implemented—can hope to succeed without the active and effective participation of the main actors—the people who must be properly educated and informed. For this age-old techniques, beliefs—mentalities must be brought in line with present day realities. People have to learn to think differently in order to veer from a course which however right in the past, has been shown to be less than adequate for present conditions—and catastrophic for the future—and must therefore needs to be altered.

Changing mentalities is neither an easy nor a rapid process. It is difficult to go back upon the accumulated experience of generations—even in the face of stark realities and scientific evidence. Moreover, when dealing with such global and fundamental issues as water, where even "scientific evidence" tends to be stated in tentative terms, the task becomes more onerous. Add to this the fact that the problem presents itself most presently in developing countries which are equally subject to enormous economic pressures which tend to reduce the scope of possible solutions. We are thus faced with the enormous task of trying to change attitudes, values, mentalities of populations whose geographical, socio-cultural and economic conditions have already fashioned priorities other than those that would precisely permit them to overcome their difficulties in a sustainable manner. In other words, of persuading people to abandon traditional short-term strategies in favour of perhaps more unattractive, but eventually sustainable, long-term practices.

A veritable Herculean labour—which can only be accomplished through information and education. And in particular, through environmental education and information whose avowed aim is precisely to develop the understanding, knowledge, skills and motivations leading to the acquisition of attitudes, values and mentalities which are necessary to deal effectively with environmental issues and problems. Sound and systematic environmental education of the people associated with concerted local, national and international action, is the only means to finding a sustainable solution to this problem. The ground has to be paved through adequate information on the subject followed by educative processes adapted to specific local conditions. For a uniform education, whether formal or non-formal, might perhaps do more harm than good as its rejection, due to its unsuitability in the light of local customs, beliefs, traditions..., might only serve to reinforce the very attitudes that it seeks to change. In each region, each country, each locality the educative processes must correspond to the socio-cultural, historical, economic conditions of the people. Only then can we hope to arrive at the change in mentalities around the planet which, coupled with consistent, parallel support from national and international institutions, will lead to the safeguard of what is perhaps our most precious resource—Water.

14

Population Growth and Fresh Water

Wherever population is growing, the supply of fresh water per person is declining. As a result of population growth, the amount of water available per person from the hydrological cycle will fall by 74 per cent between 1950 and 2050. Stated otherwise, there will be only one fourth as much fresh water per person in 2050 as there was in 1950. With water availability per person projected to decline dramatically in many countries already facing shortages, the full social effects of future water scarcity are difficult even to imagine. Indeed, spreading water scarcity may be the most underrated resource issue in the world today.

Evidence of water stress can be seen as rivers are drained dry and as water tables fall. The Colorado River in the south-western United States now rarely reaches the sea. The Yellow River, the northernmost of China's two major rivers, has run dry for a part of each year since 1985, with the dry period becoming progressively longer. In 1997, it failed to make it to the sea for 226 days. The Nile, the largest river in the Middle East, has little water left when it reaches the sea.

Water tables are now falling on every continent, including in major food-producing regions. Among those where aquifers are being depleted are the U.S. southern Great Plains, the North China Plain, which produces nearly 40 per cent of China's grain; and most of India. Wherever water tables are falling today, there will be water supply cutbacks tomorrow, as aquifers are eventually depleted.

Some 70 per cent of the water pumped from underground or diverted from rivers is used for irrigation, 20 per cent is used for industrial purposes, and 10 per cent is for residential use. Water use patterns vary widely by region. In Europe, for example, where agriculture is largely rain-fed, water withdrawals are dominated by industrial use. In Asia, in contrast, irrigation accounts for 85 per cent of all water use.

As countries press against the limits of their water supplies, the competition among sectors intensifies. The economics of water use does not favour agriculture. One thousand tons of water can be used to produce one ton of wheat worth $200 or to expand industrial output by $14,000. This ratio of 70 to 1 explains why industry almost always wins in the competition with agriculture for water.

As the growing demand for water collides with the limits of supply, countries typically satisfy rising urban and residential demands by diverting water from irrigation. They then import grain to offset the loss of irrigation water. Since it takes at least 1,000 tons to water to produce a ton to grain, importing grain becomes the most efficient way to import water. North Africa and the Middle East—a region where population growth is rapid and every country faces water shortages—has become the world's fastest growing grain import market during the 1990s. In 1997, the water required to produce the grain and other foodstuffs imported into the region was roughly equal to the annual flow of the Nile River.

In both China and India, the two countries that together dominate world's irrigated agriculture, substantial cutbacks in irrigation water supplies lie ahead. The combination of the effects of aquifer depletion in key countries such as these and the growing diversion of irrigation water to non-farm uses in many countries makes it unlikely that there will be much, if any, increase in total irrigated area over the long term. Already the irrigated area per person has been slowly declining since 1978, falling from a historical high of 0.047 hectares per person to 0.045 hectares in 1996—a drop of 4 per cent. If the total irrigated area remains at roughly 263 million hectares until 2050, this key figure will fall to 0.028 hectares per person in 2050—declining by an additional 38 per cent. Such a shrinkage will pose a formidable challenge to the world's farmers.

About a billion people will be living in countries facing absolute water scarcity by 2025. These nations do not have enough water to maintain 1990 levels of food production per person from irrigated area, even with high irrigation efficiency, and to meet the needs for domestic, industrial, and environmental purposes as well. They will have to reduce water use in agriculture in order to satisfy residential and industrial water needs. The resulting decline in domestic food production will force them to import more food, assuming it is available. Although detailed water projections by sector for each country are not available for 2050, the number of water-deprived people will be far greater than in 2025 if the world continues on the U.N. medium population trajectory. The bottomline is that if we are facing a future of water scarcity, then we are also facing future of food scarcity.

15

Fresh Water and the Environment

It is widely recognised that water is going to be one of the major issues confronting humanity at the turn of the century and beyond. We are facing a crisis as regards the quantity and quality of water supply, but we have yet to experience the full social and political impact of that crisis. The escalation in the population and the quest for continued development is leading to conflicting pressures on water resources. Such resources are the ultimate recipient of pollution from various socio-economic activities associated with urbanisation, agriculture, mining and clearing of native vegetation. Pollution originating from human waste, especially where appropriate sanitation facilities are not available, or are located too close to water supply sources affects both surface water and ground water.

This makes water supply and health perhaps the most important issues for the larger proportion of the global population. Paradoxically, the demands for "sustainable management" and increasing global population require more potable water from a declining available potable water base.

It is universally accepted that proper water administration is a critical component of sustainable development—that is, development that meets the needs of both present and future generations. Indeed, water is an essential factor in a large number of productive activities, of which one of the most important is the production of food by irrigation. This activity, accounts for two thirds of the water resources used by humanity. Supply of drinking water and sanitation in urban centres are crucial for preserving human health.

For some decades it has been known that the misuse of water resources is responsible for many important environment problems. For example, in many industrialised cities both surface water and ground water are seriously contaminated. This deterioration is a consequence of a range of human activities, sometimes in isolation, others over a large area or a long period of time. Among examples of the latter is modern agriculture, whether it uses irrigation or not, as a result of the intensive use of mineral fertilizers and pesticides.

Water Shortage: Exaggeration, Reality or Bad Management?

Some of these problems have made news and have created the impression that water shortage will be one of humanity's big problems in the coming decades. Sometimes this feeling is due to genuinely manipulative publicity campaigns to justify the setting in motion of hydraulic mega projects which basically benefit a few large construction companies. The truth is that except for a handful of very specific cases, no problems of water shortage are to be found almost anywhere. On the other hand, cases of bad water management are not rare at all.

Basic Principles for Good Water Management

Good management of water resources—and of almost all other natural resources—must be based on the principles of solidarity, 'subsidiarity' and participation. The physical reality requires that these resources be considered a common heritage of humanity both now and in the future. By "subsidiarity" we mean that water management should be as decentralised as possible: what one person or any minor social group can do should not be done by a higher authority. For example, what local government can do should not be done by a regional, state or central government. Participation consists in water users playing as large a part as possible in decisions affecting water, in keeping with each state's or country's social and cultural structure. Obviously this participation calls for a certain cultural and technical knowledge—a hydrological education—on the part of the users.

The need for participation by users is even greater in the exploitation of groundwater. In this case, users tend to extract water independently of one another. They often fail to realise, until there is a serious economic or environmental impact, that their pumping affects other people who rely on the same water supply as has happened.

Water shortage is rarely a serious problem: in fact, in some cases the problem is exaggerated to justify the construction of large works using taxpayers' money. On the other hand, the contamination of surface and groundwater tends to be a problem which rarely receives adequate treatment. Successful water management should be based on three basic principles: solidarity, subsidiarity and participation. The specific way in which these principles are applied will vary from one state or country to another, but the effectiveness of water management will depend in large measure on the hydrological education of the general public.

The universal way of obtaining freshwater is from rain. River systems are the results of the excess water that falls on dry land in the form of rain. On the one hand, rainwater penetrates the permeable soils, saturates them and accumulates to form groundwater reservoirs, or aquifers, which can come to the surface in the form of springs. On the other hand, the water is absorbed by vegetation, which uses it for pumping minerals and then evaporates it by transpiration. Some rainwater is lost because it evaporates immediately on falling on impermeable surfaces like the asphalt of roads and cities. Running water courses finally flow over saturated soils, shaping the complex systems of the watersheds or river basins.

Since each basin's natural system has developed gradually and has grown up according to the yearly distribution and fluctuations of rainfall, we have to appreciate that any large-scale project for redistributing water by means of pipes, as if it were gas or electricity, is a journey into the unknown. This is because it destroys the results of the work of shaping the climate, however transitory it might be.

Variable Volumes

All water supplies are of variable volume. Both the discharge of river and the level of lakes and aquifers depend on rainfall. As these resources are components of a larger system, the river basin, a reasonable policy would be to manage water resources according to the characteristics of each basin. This would require, first of all, a proper understanding of the system so as to adapt use and consumption of the existing supply. Conserving river systems as much as possible in their natural state is the best guarantee for the preservation of the landscape and of a constant supply. Groundwater reservoirs aren't canals, but are more like lakes, so that pollution leads to the build-up of a debt which is paid in years to come.

Consumption

Water consumption has increased in recent years as a result of not only population growth but also an increase in living standards. In the rural areas the introduction of new farming methods, the spread of irrigation and the excessive use of fertilizers and pesticides causes very high consumption—it is estimated that more than 2/3 of water consumption is used in irrigating. Agricultural pollution also endangers both surface water and aquifers, which receive water full of chemical products. Many causes of eutrophication, the enrichment of water by nutrients that accelerate the growth of algae, derive from the run-off of fertilizers. The practice of intensive stock-raising on farms with large numbers of animals also brings about these problems of over consumption and pollution. Cleaning the stockyards requires large amounts of water which is then released into the environment with high concentrations of nitrogen.

As for industries, they have in the past taken little care over water consumption and dumping, and in many areas the need for proper attention comes as something new. The best thing would be to make industry take its water at a point down-river from where it returns it or, better still, generalise the use of closed circuit systems based on the constant recycling and reusing of the same water.

As regards human consumption, the general attitude to cleanliness is based on diluting pollutants. One example is the success of the use of the Water Closet which involves diluting a few decilitres of urine in 10 or more litres of drinking water: quite a record in wastefulness.

Another aspect to be considered is the different quality of the water that falls on well formed soils from the water that falls on roads, cities, airports, suburbs and built-up areas and whose composition is less stable and "worse" than that resulting from a more uniform interaction with mature soils. Remember that streets, roofs, communication routes, airports and built-up areas already cover a high proportion of the earth's land area and are still on the increase.

Purification techniques should be based especially on the natural processes that include biological activity. Otherwise, for example, if physico-chemical methods are used there can be side-effects such as an excess of mud or sediments. The strategy to follow is to optimise operations in our use of water according to the discharge and to the distribution of contamination. A system in the form of a conduit or channel, such as a river, can respond relatively quickly. On the other hand, lakes and dams can only do so up to a point, because they show more inertia and irreversibility and take longer to clean.

Large lakes, not to mention the sea, might seem a good place to dump contaminating refuse, but they can't then be cleaned. This is the price we pass on to the future generations: a comfortable attitude, but an unacceptable one.

16

Watershed Development Programme

The Commissionerate of Rural Development is implementing area development programme, i.e., watershed development programme, in dry and degraded lands through district agencies—DDP/DPAP/DWMA.

The programme of dry land development in Andhra Pradesh has undergone a major change from 1995-96, with the introduction of new participatory watershed guidelines based on the recommendations of Dr. Hanumantha Rao Committee. The main emphasis of the revised guidelines is on active mobilisation and participation of the stakeholders in the programme at all stages—planning, implementation and management. So far, out of 78.20 lakh ha. dry land to be treated by the Rural Development Department, 36.85 lakh ha. is treated/under treatment through 7903 watersheds/projects since 1995-96. The following watershed development programmes are being implemented in Andhra Pradesh:

1. Drought Prone Area Programme (DPAP)
2. Desert Development Programme (DDP)
3. Integrated Wastelands Development Programme (IWDP)
4. Employment Assurance Scheme (EAS)
5. Andhra Pradesh Hazard Mitigation (APHM)
6. Rural Infrastructure Development Fund (RIDF).

In order to combat the frequent recurrence of drought and for development of wasteland in the state, watershed programme is being implemented. A massive action plan for the development of all the degraded and wastelands in ten years was launched during 1996-97. Ten-year action plan has been prepared to develop 100 lakh hectares of degraded and wastelands on watershed basis with an outlay of about Rs. 4000.00 crores from 1997-2007 at the rate of 10 lakh hectares every year.

Andhra Pradesh is one of the few states in the country to have an exclusive administrative agency—DPAP/DDP/DWMA for the implementation of watershed development programme.

DPAP/DDP

Since 1995-96, 2966 DPAP and 552 DDP watershed projects are under implementation in Andhra Pradesh covering an area of 14.83 lakh ha. and 2.76 lakh ha. respectively (total = 17.59 lakh ha.).

291 DPAP and 110 DDP watershed projects are under implementation during 2002-03.

During 2003-04, action was taken to cover more dry areas depending upon the projects to be sanctioned by the Government of India.

IWDP

Since 1995-96, 842 watersheds have been sanctioned up to December 2002 to develop 4.78 lakh ha. of wastelands in non-DPAP blocks under IWDP scheme.

108 watersheds have been sanctioned during 2002-03 up to December 2002 as against target of 144 watersheds under IWDP scheme.

During 2003-04, action was taken to cover more dry areas depending upon the projects to be sanctioned by the Govt. of India.

EAS (Watersheds)

Since 1995-96, 1884 watersheds have been taken up covering an area of 9.42 lakh ha. and component-wise action is being taken to complete them by March 2003.

RIDF-VI

1244 RIDF watershed projects have been sanctioned to treat an area of 2.63 lakh ha. with Rs. 139.19 crores in 22 districts and we would complete all works by March 2004. The RIDF—VI projects have been sanctioned with 90 per cent as NABARD loan and 10 per cent GOAP share.

RIDF-VIII (RD and Other Sectors)

NABARD has sanctioned 290 SMC projects under RIDF-VIII (RD & Other Sectors) to cover an area of 1.48 lakh ha. with Rs. 41.08 crores to 17 districts with 75 per cent NABARD loan of Rs. 30.81 crores and 25 per cent of project cost with rice component of Rs. 10.27 crores during 2002-03.

APHM and ECRP

It is a World Bank assisted programme which commenced in July 1997 with a target of 100 watersheds @ 20 watersheds per district. The total area treated is 0.5 lakh ha. The project has been completed in July 2002.

Neeru-Meeru Programme

The Government of Andhra Pradesh has launched Neeru-Meeru programme with a view to converge efforts of all the departments involved in water conservation and management and to ensure large scale participation of people.

The programme was initiated on 01.05.2000 coinciding with 12th round of Janmabhoomi programme. Neeru-Meeru activities taken up by different departments are aimed at creating more filling space for harvesting rainwater which contributes to additional groundwater recharge. Seven line departments (i.e., Rural Development, Forest, Minor Irrigation (I&CAD), Minor Irrigation (PR), Rural Water Supply (PR), Municipal Administration and Endowments) are actively involved in conservation and management of water resources.

Watershed Development Programme—Performance and Impact

The performance of watershed programme in Andhra Pradesh is as follows:

SCHEME-WISE WATERSHEDS

(FROM 1995-96 ONWARDS) UP TO DECEMBER 2002

S.No.	*Scheme*	*No. Water-Sheds*	*Area under Treatment in Lakh ha.*
1.	DPAP	2966	14.93
2.	DDP	552	2.76
3.	IWDP	842	4.78
4.	EAS	1884	9.52
5.	APHM	100	0.50
6.	RIDF-VI	1244	2.63
7.	RIDF-VIII	290	1.48
8.	WDF	25	0.25
	Total	**7903**	**36.85**

Watershed projects under implementation : 7903

Land under watershed treatment : 36.85 lakh ha.

Amount invested on participatory watershed programme (up to December 2002) : Rs. 992.29 crores

Amount generated as people's contribution for watershed development fund : Rs. 49.61 crores

Impact

Andhra Pradesh State Remote Sensing Application Centre (APSRAC) is using satellite application and the satellite imageries for evaluation. The Annual evaluation of watersheds in September 2002 has revealed the following impact and it shows that the programme is immensely useful to the farmers and the poor in dry areas.

1. No. of districts covered : 20
2. No. of watersheds evaluated : 5298
3. Average increase in water levels (metres) : 1.96

4.	Per cent No. of wells rejuvenated	:	43%
5.	Additional area brought under cultivation (Ac.)	:	3,34,755
6.	Decrease in labour migration	:	61%
7.	Increase in milk production (ltrs. per day)	:	3,71,328
8.	Additional area brought under horticulture/afforestation (Ac)	:	3,56,046

D.P.A.P., KURNOOL
WATERSHED DEVELOPMENT PROGRAMME

1.	Total Geographical Area	17.63	lakhs ha.
2.	No. of Mandals	53	
3.	No. of Mandals Covered	48	
4.	No. of Habitations Covered	436	
5.	Total Area Included	4.40	lakh ha.
6.	No. of Micro Watersheds of 500 Ha. each	841	EAS-355 IWDP-13 DPAP-373 APRLP-100
7.	No. of Project Implementing Agencies	71	
8.	No. of Government P.I.As	48	
9.	No. of Non-Government P.I.As	23	
10.	No. of Watersheds Registered	841	
11.	No. of Watershed Bank Accts. opened	841	
12.	No. of User Groups	6071	
13.	No. of Self Help Groups	4095	

D.P.A.P., KURNOOL
PHYSICAL AND FINANCIAL PERFORMANCE FOR 2002-03

Soil and Moisture Conservation

Sl. No.	Name of the Scheme	Annual Target		Target up to December '02		Achievement up to December '02		Balance	
		Phy.	Fin.	Phy.	Fin	Phy.	Fin.	Phy.	Fin.
1.	DPAP	3053	67.863	2291	39.683	2048	28.372	762	28.181
2.	EAS	5114	113.687	3838	66.477	3317	45.938	1276	47.209
3.	RIDF-VI								
4.	RIDF-VIII								
	Total	**8167**	**181.550**	**6129**	**106.160**	**5365**	**74.310**	**2038**	**75.390**

(Contd...)

Contd...

Sl. No.	Name of the Scheme	Water Harvesting Structures							
		Annual Target		Target up to December '02		Achievement up to December '02		Balance	
		Phy.	*Fin.*	*Phy.*	*Fin*	*Phy.*	*Fin.*	*Phy.*	*Fin.*
1.	DPAP	5310	457.154	3983	368.335	11349	442.010	1327	88.819
2.	EAS	7315	286.926	5255	283.225	16233	360.558	2060	27.521
3.	RIDF-VI	568	425.780	406	280.690	1012	304.510	162	121.270
4.	RIDF-VIII	1012	53.130	1012	53.130	1132	50.622	0	0.000
	Total	14205	1222.990	10656	985.380	29726	1157.700	3549	237.610

(Contd...)

Contd...

Sl. No.	Name of the Scheme	Afforestation and Horticulture							
		Annual Target		Target up to December '02		Achievement up to December '02		Balance	
		Phy.	Fin.	Phy.	Fin.	Phy.	Fin.	Phy.	Fin.
1.	DPAP	3686	197.108	3686	118.965	4995	96.025	0	78.143
2.	EAS	5725	272.351	5725	141.443	7639	97.631	0	130.907
3.	RIDF-VI	450	57.851	450	57.851	450	57.851	0	0.000
4.	RIDF-VIII								
	Total	**9861**	**527.310**	**9861**	**318.259**	**13084**	**251.507**	**0**	**209.050**

(Contd...)

Contd...

Trainings, Comm. Org. and Est.

Sl. No.	*Name of the Scheme*	*Annual Target*	*Target. up to December '02*	*Achievement up to December '02*	*Balance*	*Annual Target*	*Target up to December '02*	*Achievement up to December '02*	*Balance*
1.	DPAP	114.794	86.094	62.951	28.700	836.919	613.077	629.358	223.843
2.	EAS	192.306	144.226	101.926	48.080	865.270	635.371	606.053	253.717
3.	RIDF-VI					483.631	338.541	362.361	121.270
4.	RIDF-VIII					53.130	53.130	50.622	0.000
	Total	**307.100**	**230.320**	**164.877**	**76.780**	**2238.950**	**1640.119**	**1648.394**	**598.830**

WATERSHED DEVELOPMENT PROGRAMME: KURNOOL DISTRICT

	Total Watersheds	841
(a)	Completed	260
(b)	Ongoing	581
	Mandals covered	48
	Villages covered	436
	Area covered (lakh ha.)	4.40

17

A Rare and Precious Resource

Fresh water is a scarce commodity. Since it's impossible to increase supply, demand and waste must be reduced. But how?

Water is a bond between human beings and nature. It is ever-present in our daily lives and in our imaginations. Since the beginning of time, it has shaped extraordinary social institutions, and access of it has provoked many conflicts.

But most of the world's people, who have never gone short of water, take its availability for granted. Industrialists, farmers and ordinary consumers blithely go on wasting it. These days, though, supplies are diminishing while demand is soaring. Everyone knows that the time has come for attitudes to change.

Few people are aware of the true extent of fresh water scarcity. Many are fooled by the huge expanses of blue that feature on maps of the world. They do not know that 97.5 per cent of the planet's water is salty—and that most of the world's fresh water—the remaining 2.5 per cent—is unusable: 70 per cent of it is frozen in the icecaps of Antarctica and Greenland and almost all the rest exists in the form of soil humidity or in water tables which are too deep to be tapped. In all, barely one per cent of fresh water — 0.007 per cent of all the water in the world, is easily accessible.

Over the past century, population growth and human activity have caused this precious resource to dwindle. Between 1900 and 1995, world demand for water increased more than six-fold compared with a three-fold increase in world population. The ratio between the stock of fresh water and world population seems to show that in overall terms there is enough water to go round. But in the most vulnerable regions, an estimated 460 million people (8 per cent of the world's population) are short of water, and another quarter of the planet's inhabitants are heading for the same fate. Experts say that if nothing is done, two-thirds of humanity will suffer from a moderate to severe lack of water by the year 2025.

Inequalities in the availability of water—sometimes even within a single country—are reflected in huge differences in consumption levels.

Scarcity is just one part of the problem. Water quality is also declining alarmingly. In some areas, contamination levels are so high that water can no longer be used even for industrial purposes. There are many reasons for this—untreated sewage, chemical waste, fuel leakages, dumped garbage, contamination of soil by chemicals used by farmers. The worldwide extent of such pollution is hard to assess because data are lacking for several countries. But some figures give an idea of the problem. It is thought for example that 90 per cent of waste water in developing countries is released without any kind of treatment.

Things are especially bad in cities, where water demand is exploding. For the first time in human history, there will soon be moré people living in cities than in the countryside and so water consumption will continue to increase. Soaring urbanisation will sharpen the rivalry between the different kinds of water users.

Curbing the Explosion in Demand

Today, farming uses 69 per cent of the water consumed in the world, industry 23 per cent and households 8 per cent. In developing countries, agriculture uses as much as 80 per cent. The needs of city-dwellers, industry and tourists are expected to increase rapidly, at least as much as the need to produce more farm products to feed the planet. The problem of increasing water supply has long been seen as a technical one, calling for technical solutions such as building more dams and desalination plants. Wild ideas towing chunks of icebergs from the poles have even been mooted.

But today, technical solutions are reaching their limits. Economic and socio-ecological arguments are levelled against building new dams. For example, dams are costing more and more because the best sites have already been used, and they take millions of people out of their environment and upset ecosystems. As a result, twice as many dams were built on average between 1951 and 1977 than during the past decade.

Hydrologists and engineers have less and less room for manoeuvre, but a new consensus with new actors is taking shape. Since supply can no longer be expanded—or only at prohibitive cost for many countries—the explosion in demand must be curbed along with wasteful practices. An estimated 60 per cent of the water used in irrigation is lost through inefficient systems, for example.

Economists have plunged into the debate on water and made quite a few waves. To obtain "rational use" of water, i.e. avoiding waste and maintaining quality, they say consumers must be made to pay for it. Out of the question, reply those in favour of free water, which some cultures regard as "a gift from heaven". And what about the poor, ask the champions of human rights and the right to water? Other important and prickly questions being asked by decision-makers are how to calculate the "real price" of water and who should organise its sale.

The State as Mediator

The principle of free water is being challenged. For many people, water has become a commodity to be bought and sold. But management of this shared resource cannot be left exclusively to market forces. Many elements of civil society—NGO's, researchers, community groups—are campaigning for the cultural and social aspects of water management to be taken into account.

Even the World Bank, the main advocate of water privatisation, is cautious on this point. It recognises the value of the partnerships between the public and private sectors which have sprung up in recent years. Only the state seems to be in a position to ensure that practices are fair and to mediate between the parties involved—consumer groups, private firms and public bodies. At any rate, water regulation and management systems need to be based on other than purely financial criteria. If they aren't, hundreds of millions of people will have no access to it.

18

WTO Agricultural Negotiations:

Completing the Task

The Cairns Group of 15 agricultural-exporting countries was formed in 1986 to influence agricultural negotiations within the World Trade Organication (WTO). It was largely as a result of the group's efforts that a framework for reform in farm products trade was established in the Uruguay Round and agriculture was for the first time subject to global trade liberalising rules. The group is positioning itself to play an important role in the new round of WTO agricultural negotiations.

The Cairns Group, which accounts for about 20 per cent of world agricultural exports, includes both developed and developing countries across a diverse set of regions around the world. The group consists of Argentina, Australia, Brazil, Canada, Chile, Colombia, Fiji, Indonesia, Malaysia, New Zealand, Paraguay, Philippines, South Africa, Thailand and Uruguay. By acting collectively, this disparate group has had more influence and impact on the agriculture negotiations than individual members would have had independently. Under Australian leadership, the group takes a consensual approach to decision-making.

Beyond the Uruguay Round

Members of the Cairns Group were generally pleased with the Uruguay Round outcome, but believe much remains to be done to ensure that a genuine market-oriented approach to agricultural policies is achieved. For example, in 1997 levels of agricultural support in Organisation for Economic Cooperation and Development (OECD) countries alone were still extremely high at $280 billion. The approach taken by the group to the challenge of reducing this assistance and creating a freer agricultural marketplace has been in two parts. First, the group has worked to ensure that countries meet the commitments that were agreed to in the agricultural-related agreements during the Uruguay Round. It has done this by remaining visible and active since the end of the round.

Second, the Cairns Group has been effective in engaging other WTO member countries in early preparation for the next round of agricultural negotiations in an attempt to ensure that they start on time and are not unnecessarily protracted as they were during the Uruguay Round. The Cairns Group in April 1998 agreed on a strongly worded "vision statement" conveying the Group's ambition and broad objectives for the 1999 agriculture negotiations and initiated a strategic approach to the preparations for the negotiations. This approach is necessarily ambitious: "The Cairns Group of Agricultural Fair Traders reaffirms its commitment to achieving a fair and market-oriented agricultural trading system as sought by the Agreement on Agriculture. To this end, the Cairns Group is united in its resolve to ensure that the next WTO agriculture negotiations achieve fundamental reform which will put trade in agricultural goods on the same basis as trade in other goods. All trade-distorting subsidies must be eliminated and market access must be substantially improved so that agricultural trade can proceed on the basis of market forces."

Objectives for Negotiations

The vision statement outlines the Cairns Group's reform goals in three key areas within the Uruguay Round framework, as follows:

- Deep cuts to all tariffs are required, as well as the removal of tariffs peaks and the redressing of tariff escalation so that market access for agricultural commodities and value-added agricultural products is on a similar footing as trade in other commercially traded products. This should include the objective of transforming market access barriers to tariffs and removal of nontariff barriers to trade. In the interim, the Cairns Group supports substantial increases in trade volumes under tariff-rate quotas, while the administration of tariff-rate quotas must not diminish the size and value of market access opportunities, particularly in products of special interest to developing countries.

- All trade-distorting domestic supports must be eliminated or replaced with non-trade distorting methods of assistance. Income aids or other domestic support measures should be targeted, transparent, and fully decoupled so that they do not distort production and trade.
- Export subsidies must be made illegal for agricultural products, as they are for other traded goods, and clear rules must be established to prevent circumvention of export subsidy commitments. In this regard, it is worth noting that only 25 of the 134 current WTO members are entitled to use export subsidies, and most of these are developed countries (with more than 80 per cent of export subsidies accounted for by the European Union). Also, agricultural export credits must be brought under effective international discipline with a view to ending government subsidisation of such credits.

Special Needs of Developing Countries

The vision statement also reaffirms the group's support for the principle of special and differential treatment for developing countries, including least-developed countries and small states, remaining an integral part of the next WTO agriculture negotiations. The Cairns Group ministers agreed that the framework for liberalisation must continue to support the economic development needs, including technical assistance requirements, of these WTO members. As has been stated by the Cairns Group: Major challenges facing many developing countries are the persistence of rural poverty and the linkages between such poverty and serious environmental problems. Consequently, more sustainable agricultural development remains a central policy issue in many developing countries. An improved international trading environment that is more conducive to supporting agricultural development is needed as an essential ingredient in addressing these problems.

Adherence to these principles will not only improve the trading environment for agricultural exporting nations, but will also have important implications for global food security. Food security will be enhanced through more diversified and reliable sources of supply, as more farmers, including poorer farmers in developing countries, are able to respond to market forces and new income-generating opportunities, without the burden of competition from heavily subsidised products. To provide further assurance to net-food-importing countries, export restrictions must not be allowed to disrupt the supply of food to world markets.

Reductions in assistance to the agricultural sector may also have positive implications for the environment. In many cases, agricultural subsidies and access restrictions have stimulated farm practices that are harmful to the environment. Reform of these policies can contribute to the development of environmentally sustainable agriculture.

Preparations for the Next Round

Cairns Group ministers welcomed the launch by the second WTO Ministerial Conference in Geneva in May 1998 of preparations for the next round of agriculture negotiations. The WTO Ministerial Declaration that emanated from this conference binds WTO members to a preparatory process that began in September 1998 and will culminate in ministerial agreement on a decision on the scope, structure, and time-frame for the agriculture negotiations.

The Cairns Group reaffirms its commitment to achieving a fair and market-oriented agricultural trading system as sought by the Agreement on Agriculture. To this end, the Cairns Group is united in its resolve to ensure that the next WTO agriculture negotiations achieve fundamental reform that will place trade in agricultural goods on the same basis as trade in other goods.

The Uruguay Round Agreement on Agriculture

The Uruguay Round Agreement on Agriculture (URAA) calls for the initiation of negotiations for continuing the process of agricultural trade reform in 1999. Article 20 of the agreement states that member countries of the World Trade Organisation (WTO) recognise that the long-term objective of substantial progressive reductions in trade distorting support and protection of agriculture resulting in fundamental reforms is an ongoing process.

The Uruguay Round Agreement on Agriculture, which entered into force in 1995 along with other Uruguay Round accords, including the agreement to establish the World Trade Organisation, was an important step toward applying multilateral rules and disciplines to global agricultural trade. Most assessments of the agreement hail it as a historic shift in the way agriculture establishes new multilateral trade agreements. The agreement establishes new multilateral rules governing market access, export subsidies and domestic support for agriculture. In terms of future trade liberalisation, its most important provisions may be those requiring the elimination of quantitative trade restrictions and their conversion to bound tariffs. These bound tariffs, even if some of them are extremely high, can provide a starting point for future negotiations of tariff reductions.

Market Access

The agreement requires all WTO members to convert nontariff trade barriers to tariffs and to reduce them by a simple average of 36 per cent over six years (with a minimum tariff reduction per tariff line of 15 per cent). The agreement prohibits the introduction of new nontariff barriers to trade. Where nontariff barriers restrict imports, the agreement requires that importing countries offer minimum access of usually 3 per cent of consumption rising to 5 per cent over the six-year implementation period for the agreement.

Most assessments of the agreement conclude that it provides little in the way of expanded access for agricultural products. Its importance lies in extending the principle (already applied to trade in industrial products) of protection by bound tariffs to agricultural trade and establishing at least a base for further tariff reductions in future negotiations.

Export Subsidies

The agreement requires that export subsidies be reduced by 21 per cent in terms of quantities and by 36 per cent in terms of budgetary outlays by the end of the six-year implementation period. WTO members may continue to use their existing export subsidies within the limits established, but may not introduce any new export subsidies.

Domestic Support

The agreement also includes rules and commitments for domestic support. Domestic subsidies are to be cut by 20 per cent from average levels of support aggregated across all commodities for the base period 1986-88. Support reduction commitments are also to be made over the six-year implementation period on the basis of this aggregate measure of support (AMS).

Trade policy experts contend that the rules established for domestic support policies are more important than the reduction commitments required. The agreement defines which domestic policies are permitted ("green box" policies), such as income support provided to farmers independently of participation in production-limiting programmes, advisory services, or domestic food assistance. Policies that are not eligible for the green box are automatically prohibited ("amber box" policies.)

Sanitary and Phytosanitary Measures

An agreement on the Application of Sanitary and Phytosanitary (SPS) Measures reaffirms the right of WTO members to adopt and enforce measures that they deem appropriate to protect human, animal, or plant life or health as long as such measures are not applied in an "arbitrary and unjustified" manner. The agreement states that such measures may not be used as disguised barriers to trade. SPS measures may be based on international standards where they exist. WTO members could impose higher standards than those derived from these sources if based on scientific justification and risk assessment. All WTO members agree to recognise the equivalence of different standards that result in a comparable level of SPS protection. Dispute settlement panels should seek advice from relevant international organisations when scientific or technical matters are at issue.

The SPS Agreement, though binding on WTO members, is stated in broad language. Specifics will come from interpretation of the agreement and adjudication of sanitary and phytosanitary issues in WTO dispute settlement.

Dispute Settlement

New and strengthened dispute settlement procedures agreed to as part of the Uruguay Round also apply to disputes that may arise under either the Agreement on Agriculture or the SPS Agreement. An important change in WTO dispute settlement procedures is the elimination of a member's right to veto a dispute panel's decision and effectively block implementation of the Panel's recommendations for resolving the dispute. Potentially this strengthens the ability of the WTO to enforce panel judgements. The right of WTO members to negotiate compensation rather than change its challenged policies remains in place, however.

20

Developing Countries and The WTO Agricultural Negotiations

Developing countries as a group have much to gain from continued progress toward a transparent, rule-based trading system in agriculture. The researchers say the negotiations should eliminate export subsidies, impose stricter disciplines on export taxes, cut tariffs, and ensure that food aid continues to be available to poor countries in grant form and delivered so as not to displace domestic production in the countries receiving it. Badly managed food aid, or cheap food imports due to export subsidies, may just reinforce the bias of economic policies against the rural sector. With its negative impact on poor agricultural producers, they say international research organisations (such as IFPRI, among other institutions) may provide support to developing countries through programmes of collaborative research, technical assistance, and capacity strengthening.

Starting with the first round of trade negotiations under the General Agreement on Tariffs and Trade (GATT) after World War II, there has been a relatively steady trend of increasing multilateral trade liberalisation. The successive rounds of negotiations recognised the greater needs of developing countries, especially since the Tokyo Round. Yet the participation of developing countries was limited. Since many developing countries were not members of GATT, the major forum for airing their views was provided by the United Nations Conference on Trade and Development. The views of developing countries had some impact on the Lome agreements and on aid flows, but had limited influence on negotiations concerning trading rules, which were discussed within the framework of the GATT, where OECD (Organisation for Economic Cooperation and Development) countries set the agenda.

In the Uruguay Round, which began in 1986 and concluded in 1993, developing countries played a larger role in the negotiations compared to previous rounds. In particular, agricultural net exporters organised the Cairns Group (which in addition to Australia, New Zealand and Canada, included several large developing countries such as Argentina, Brazil, Indonesia, and the Philippines) to pursue their interests. Furthermore, during and after the conclusion of the Uruguay Round, the formal accession of developing countries to the GATT and now the World Trade Organisation (WTO) has continued apace. Of the 134 members of the WTO in February 1999, some 70 per cent were developing countries. The United Nations classified 48 countries as least-developed (LLDCs). Within that group, 29 are members of the WTO, six are in the process of accession, and three are observers. Also, 18 countries have been identified as net-food-importing developing countries (NFIDCs).

Some Definitions

The LLDCs are identified by the United Nations General Assembly based on several criteria—income per capita, augmented physical quality of life index, and an index of economic diversification. As a group, they have a population of about 590 million people, with an income per capita about 4 per cent that of the world average (1996). Agricultural production per capita in LLDCs has been declining since the 1970s although the same indicator for all developing countries (mainly under the influence of China) has gone up by nearly 40 per cent in the same period. LLDCs represent a small fraction of world trade (less than 1 per cent for total and about 2 per cent for agricultural trade). They had a positive, although declining net agricultural trade balance until the mid-1980s, when it turned negative. Almost 20 per cent of their total imports are food items.

The 18 net-food-importing developing countries have been selected through a process within the WTO. They have a population of some 380 million people and an income per capita nearly five times that of the LLDC average, but still much lower than the world average. NFIDCs are a diverse group: four are upper-middle income countries; eight are lower-middle income; and six are lower income. Four of them had net food exports on average during 1995-97, but because they imported cereals they are included in the group. NFIDCs' per capita food production as share of both world and developing country averages has risen, although from very low levels.

Although the categories of "developed" and "developing" countries have important legal consequences under WTO rules, there are no formal definitions of either category. The process works through self-identification and negotiation with other member countries of the WTO.

Completing the Unfinished Agenda

In general, developing countries operate under what has been called "special and differential treatment". They face lower disciplines and enjoy longer time frames for implementing reforms. In the case of LLDCs, they are totally exempted from WTO commitments, and it has been agreed that developing and least-developed countries should receive special consideration for market access and technical and financial support. Also, during the Uruguay Round, concerns that liberalisation of agricultural policies and trade could adversely affect the food imports of LLDCs and NFIDCs led participants to include several measures dealing with food security issues in the "green box" of permitted domestic support—for instance, the formation of public stockholding and the provision of foodstuffs at subsidised prices. There was a ministerial decision in Marrakesh in April 1994 to deal with possible negative effects of agricultural trade reforms on the food security of LLDCs and NFIDCs. The decision was reemphasized at the 1996 ministerial meeting of the WTO in Singapore.

Export and Domestic Subsidies

While many developing countries have significantly reduced distorting domestic agricultural policies, the possible benefits that these countries and the world can enjoy are thwarted by the subsidies of developed countries. The Uruguay Round was a first step in imposing discipline on the unfair competition arising from subsidised agricultural exports, which hurts poor agricultural producers in developing countries irrespective of their net agricultural trade position. In the next negotiations, that first step should be completed with the elimination of export subsidies. Net-food-importing developing countries should also be interested in stricter disciplines on export taxes and controls that exacerbate price fluctuations in world markets.

Under the Uruguay Round agreement, there is still a lot of scope for the developed countries to use domestic subsidies, in addition to the use of export subsidies; to help their farmers. The developing countries should seek further disciplines in this regard, including, among other things, the elimination of exemptions under the "blue box" (which allows farmers to receive some forms of direct payments that are considered to be trade distorting). Least-developed and developing countries, however, will still be allowed "special and differential treatment" on these issues.

Market Access

If the developing countries are to succeed in diversifying their agricultural sectors, they need expanded access to markets in developed countries. This includes increasing the volume of imports allowed under the current regime of tariff-rate quotas (TRQs, which replaced the previous system of rigid quotas with a combination of a quantitative quota and a high tariff for the eventual out-of-quota imports); making the administration of the TRQs more transparent and equitable; seeking further reductions in tariffs, particularly those still high in some key products; and completing the process of tariffication in the cases where exemptions were granted. Also, eliminating, or at least reducing, tariffs escalation in non-agricultural products is important for developing countries: this practice undermines the possibilities of expanding production and exports of processed goods that use agricultural inputs, exploiting "forward linkages" in the value-added chain.

What the Most Vulnerable Need?

The special situation and concerns of least-developed countries are net-food-importing countries were recognised in a ministerial decision agreed upon at the completion of the Uruguay Round in 1993. These concerns include the preservation of adequate levels of food aid, the provision of technical assistance and financial support to develop the agricultural sector in those countries, and the continuation and expansion of financial facilities to help with structural adjustment and short-term difficulties in financing food imports. It is important to make food aid available in grant form, to target it to poor countries and social groups, and to deliver it in ways that do not displace domestic production in the countries receiving it. Badly managed food aid, or cheap food imports due to export subsidies, may just reinforce the bias of economic policies against the rural sector, with its negative impact on poor agricultural producers.

Volatility in agricultural prices must be monitored carefully. While expansion of world agricultural trade should limit overall fluctuations by spreading supply and demand stocks over larger areas, the decline in world public stocks as a percentage of consumption works in the opposite direction. Improving early warning of potential food shortages, lowering costs for food transportation and storage, and providing better targeted food aid programmes and financial facilities for emergencies are also issues that need to be addressed by countries participating in the coming round of negotiations.

The impact of changes in trade and agricultural policy on poorer consumers and producers in developing countries is a matter of debate. Some have argued that trade liberalisation may hurt both groups. Others have answered that greater productivity and growth coming from better trade and sectoral policies should help generate employment and income, given a setting of adequate overall economic policies and properly functioning markets and social institutions.

Small producers will also be helped by the disciplines that the URAA is bringing to subsidised and dumped exports, while it allows the implementation of a variety of programmes aimed at poor producers or consumers, including stocks for food security purposes and domestic food aid for populations in need. The issue here is the adequate design and funding of domestic policies to achieve the intended objectives of agricultural growth and poverty alleviation, which most certainly will not be helped by trade-distorting interventions either in developed or developing countries.

In general, low-income developing countries and LLDCs should emphasize to the international community the importance of creating and expanding a supportive international trade and financial environment and of implementing an integrated framework for economic and social development with agricultural and trade policies being an integral part of the strategy. Appropriate measures would include—in addition to the agricultural trade issues suggested here—the continuation and enhancement of the reduction of the external debt of Heavily Indebted Poor Countries (The HIPC initiative) and the further liberalisation of trade in textiles.

But improved international conditions should go hand-in-hand with a better domestic framework in developing and least-developed countries, including stable macroeconomic policies, open and effective markets, good governance, the rule of law, a vibrant civil society, and programmes and investments that expand opportunities for all, with special consideration for poor and disadvantaged groups.

Bringing Developing Countries into the Process

Developing countries, as small players in the global arena, should be interested and active participants in the design and implementation of international rules that limit the ability of larger countries to resort to unilateral action. Also, domestic, legal and institutional frameworks in developing countries may be strengthened by the implementation of internationally negotiated rules that limit the scope for rent seeking and arbitrary projectionist measures. The developing countries as a group have much to gain from continued progress toward a transparent, rule-based, trading system in agriculture.

What are the requirements and skills for the developing countries to become effective members in the next WTO round? Any negotiation requires careful consideration of the legal, economic, and political dimensions that define the substance and possible evolution of the negotiations, as well as the diplomatic and negotiating techniques that may help in the attainment of the expected outcomes. Questions that need to be addressed include:

- What are the economic and social consequences of different WTO scenarios (quantitative estimation of impacts)? Knowing the impacts of alternative scenarios is crucial if developing countries are to represent their interests in the negotiation process.
- What are the legal issues being discussed (definition of obligations, exemptions, time frame, and so on)? Detailed knowledge of international trade law is crucial if developing countries are not to be "shortchanged." The devil is in the details.
- Looking at the political process, who are the main actors and their interests and what type of alliances may drive the negotiations? Negotiators must understand the political economy of their own country and of other countries in the WTO if they are to negotiate effectively.
- With these elements, an adequate diplomatic and negotiating strategy must be defined and implemented.

Developing countries that have carefully considered all four components will be better prepared to participate effectively in the coming negotiations. Of course, limited financial and human resources act as an important constraint. However, developing countries may overcome some of the problems through collective action, for instance considering the creation of alliances with respect to their main export and import commodities and the markets they approach for their exports. An example is the Cairns Group. This approach could reduce the fixed costs of negotiations. Spreading them over groups of countries, allow a better use of scarce technical expertise, and improve the bargaining position of developing countries. It could also be in the interest of the OECD countries to deal with negotiating blocs, which represent a smaller number of negotiating positions, rather than with numerous separate countries. The negotiations would be much more efficient and balanced.

21

The Future of Agricultural Trade

In the Uruguay Round, countries recognised that the long-term solution for agriculture did not lie in administered prices, trade restrictions, supply controls, and export subsidies but rather in open, non-distorted markets. It is the time to take bold steps toward bringing agricultural trade into the 21st century by accelerating agricultural trade reform.

There are four key areas for accelerating reforms: eliminating export subsidies; increasing market access though substantial tariff cuts and expansion of tariff-rate quotas; cutting further trade-distorting domestic subsidies; and ensuring technical standards are based on sound science.

The world's farmers and ranchers are facing two difficult challenges at the dawn of the 21st century. First, they are being asked to provide more products at lower cost, higher quality, greater variety, and in a safer manner than ever demanded before. Second, they are being asked to produce this abundance on a shrinking natural resources base that is often subject to government regulations. Meeting these global challenges will require unleashing the production potential of world agriculture while practising proper environmental stewardship. The ingenuity and hardwork we usually associate with farmers will be essential to meet these challenges, but they will not be sufficient unless we further reform agricultural trade to create an environment that rewards risk and investment and encourages efficiencies.

Today's Agricultural Challenges

Farmers are responsible for feeding a rapidly growing

world population. And despite progress over the years, too many people still are not getting enough food. Many countries including the United States, are working vigorously to promote technological innovations to meet the need for food and fiber in the coming years. However, as important as this work is, it is only part of the solution. These technologies and the hard work of the world's farmers need a trading environment that encourages investment and efficient production, and generates economic growth to finance production and consumption needs. Long-term trends in agriculture pose serious challenges for all farmers. The same technological advances that increase yields may result in lower prices. Increasing social concerns about effect of agricultural production on the environment and living conditions result in new restrictions on farm activities. As urban dwellers and industry stake competing claims for land, water, and energy, many producers find their ability to farm made ever more difficult.

Two approaches to organising the agricultural economy present a stark contrast in dealing with these challenges. One model, popular in Europe and Asia, is to retain an inward-looking agricultural system focused on supply control and government regulation geared to keeping farm prices high and, since guaranteed high prices are a drain on the treasury, to controlling production. Under this approach, bureaucrats try to assess the optimal level of national production—not so little that imports are needed and not so much that excess production; must be bought at high prices and then dumped on world markets. This "command-and-control" structure stifles farmer efficiency and ingenuity and distorts world markets, especially as subsidised surpluses are regularly exported; and it does not address the challenge to farmers to produce food for the next century. It also ignores the interest of domestic consumers (who have to pay high internal prices) and producers in other countries (who have to compete with subsidised products). Of biggest concern is that the anti-market policies of this approach hamstring the agriculture sector from pursuing the technological advances needed to meet its future challenges.

Another approach is to place agriculture on a more market-oriented basis, particularly by removing trade barriers and reducing trade-distorting policies. Greater market orientation was the principle that actions agreed to in the last set of multilateral trade negotiations. In the Uruguay Round, countries recognised that the long-term solution for agriculture did not lie in administered prices, trade restrictions, supply controls, and export subsidies, but rather in open, nondistorted markets. Now is the time to take bold steps toward bringing agricultural trade into the 21st century by accelerating agricultural trade reform.

The Gains From Trade

The benefit from free and fair trading of agricultural products have immediate effects on people. Eliminating trade barriers and reducing unfair competition will help ensure that farmers have incentives to produce and consumers have access to the products they desire. Liberalising agricultural trade will contribute to better resource allocation by farmers, which has conservation benefits, rewards low-cost producers, encourages efficiencies, and removes the drag on economic growth.

Opening trading opportunities also increases the food security of food-importing countries by giving supplier countries the confidence required to put more land into production and to create marketing relationships. Trade provides consumers with year-round access to a greater variety of less expensive products, while rewarding producers who are able to find and meet specific consumer demands for high-value products. In a broader context, by allowing imports that are more efficiently produced elsewhere, trade encourages specialisation in efficient agricultural and nonagricultural production.

More dramatically, trade literally saves lives. Without the international flow of food products from areas with abundant production to areas where food is scarce, many people in the world would be eating less or not at all. Trade has dynamic effects, as well, that push long-term productivity growth. For example, access to customers in overseas markets creates an incentive for technological innovation, resulting in exciting developments in improved seed varieties and production techniques. International markets also expand market outlets, raising prices and giving producers increased confidence to produce more than required merely for national needs, allowing productive farmers to not only feed their neighbours but literally feed the world.

Equally important, trade in agricultural products is becoming increasingly critical to farm and ranch incomes. Increased productivity and often times flat domestic demand increases the importance of reliable international markets. Foreign markets are not just a dumping ground for surplus products; overseas consumers value choice and quality, particularly when producers in their own country cannot meet their demands or when they are charged inflated prices. Consequently foreign and value-added agricultural producers, raising farm-gate prices and helping support the range of agriculture-related industries.

Political reality also encourages a focus on international markets: policies based on high government guaranteed prices are ultimately politically untenable because they are hugely expensive, unresponsive to the needs of customers and producers, insensitive to environmental and agronomic realities, and a shameful waste of economic assets. Rather than farming government programmes, our producers are looking for customers around the world.

While agricultural trade benefits consumers and producers alike, it is an area in which progressive reform is ardently opposed by entrenched domestic interests. Producers in some countries, cosseted by high guaranteed prices and protective tariffs, oppose any move toward greater market orientation. Intervention in the agricultural economy—measured by the Organisation for Economic Cooperation and Development by summing price supports, direct payments, and other supports as a per cent of total agricultural production—has actually increased in some countries from the levels at the beginning of the Uruguay Round. In the last set of multilateral trade negotiations, countries began the process of dismantling protection and delinking farm support from production decisions. Consequently, reforms have been undertaken by some countries.

The WTO Opportunity

The major objective in the upcoming farm talks is to accelerate the reform process initiated in the Uruguay Round. That means further substantial negotiations on tariffs, subsidies, and other trade-distorting measures so that the level and direction of trade are determined by market forces, not government intervention. Four key areas are outlined below.

(i) **Export Competition:** Export subsidies are the most distorting trade tool because the level and direction of trade is directly determined by government subsidies. Today, the European Union (EU) is the only substantial export subsidiser—nearly all other countries agreed not to use, or have only limited recourse to use, export subsidies in the last round of negotiations. EU farmers, responding to domestic prices frequently twice the world price, produce more products than can be consumed in Europe, but at such high prices that they can be sold abroad only with generous subsidies. These subsidies push other competitive suppliers out of the market (which is expensive and unfair) and discourage production in countries that have a comparative advantage in agricultural production (which is wasteful and is threatening both to the environment and to future farm production needs).

In the Uruguay Round negotiations, countries acknowledged the corrosive nature of subsidies and agreed to cap and reduce their use. The upcoming negotiations should eliminate them to ensure that countries do not resort to other policy tools that allow government spending to determine winners in the marketplace. Specifically, WTO members should look closely at curbing distorting state trading agricultural export monopolies that can disguise subsidies and exert distorting market power, along with other policies used to dispose of surplus commodities on a nonmarket basis.

(ii) **Market Access:** Measures applied at the border to stop trade currently are the principal barrier to a freer and more open trading environment for agriculture. Market access barriers deny efficient producers the chance to compete in other markets and limit the variety and quality of products available to consumers. Opening markets and maximising trade opportunities as fundamental principles of WTO, and we still have a long way to go in agriculture to open markets to competition.

The Uruguay Round Agreement set agricultural trade on a more predictable basis by requiring that all non-tariff measures, such as quotas and import bans, be converted to simple tariffs. While this was a necessary first step to removing trade barriers, many of the tariffs are still prohibitively high. For example, while the average tariff assessed by the United States on agricultural products is less than 5 per cent (and nearly zero for industrial products), the average agriculture tariff-rate quota (TRQ) where only specific quantities of imports receive low duties, many other commodities also are subject to high tariffs.

As we start the next century, higher tariffs should not stop the flow of imported agricultural products. Where TRQs remain as a transitional step before we achieve more open trade, we expect more specific disciplines on the way in which they are administered. Similarly, we need to take a hard look at agricultural state trading monopoly. Importers' use of these state traders may have been justifiable when more restrictions allowed on farm trade, but in the tariff-only regime it is hard to see why a government needs to insert itself between an export and an end-user.

(iii) Domestic Subsidies: Domestic subsidy programmes are often the root cause of other-distorting policies. Subsidy policies that increase domestic prices above world price levels can be maintained only if price-competitive imports are restricted. Additionally, overproduction generated by high domestic prices can be sold on world markets only with export subsidies that bring the price down to the world price. While reining in distortive domestic subsidy programmes has value in its own right for rationalising agricultural production, the WTO negotiations will focus on their trade-distorting elements.

In the Uruguay Round negotiations, countries agreed to distinguish trade-distorting subsidies (generally those linked to the production of a specific crop or related to price supports) from non-trade distorting subsidies (such as research and development, training and environmental production). The trade-distorting subsidies were capped, and the process of reducing allowable levels of subsidies began. This distinction is a good one: the nasty sort of subsidy that distorts markets and straitjacket producers should be cut, while programmes that will increase a country's ability to produce agricultural products in the next century without distorting production incentives should not be reduced.

(iv) **Standards:** As WTO members make progress on cutting tariffs and subsidies, the temptation increase to disguise trade barriers as health and safety measures or other innocuous-sounding "technical standards." Moreover, when regulations purportedly designed to protect health are instead vehicles for domestic protectionism, the credibility of the entire safety apparatus of a country is put up for questioning. When good science is replaced by politics, the basis for sound health policy is undermined. Therefore, increasing government accountability by putting the emphasis on sound science for health standards should discipline disguised barriers to trade and strengthen health policy.

In the Uruguay Round, countries agreed to a set of sound principles: each has the right to maintain health and safety measures, but these must be based on sound science, backed by scientific evidence and an assessment of the risk, and be no more trade-restrictive than required to meet health goals. In practice, countries have found that these principles work well—bogus measures adopted without scientific basis have been successfully challenged in the WTO without sacrificing health concerns. Creating a supportive environment for the propagation of yield-enhancing biotech products also is critical for meeting the needs of the coming century.

Agriculture is Different

Agriculture occupies a special place in the national economies of most countries around the world. Farmers are responsible for feeding and clothing people. Farming also holds a powerful claim on our national cultures that calls for the preservation of rural lifestyles and values. Farm production is subject to the cruel vagaries of whether and the relentless decline in prices and increases in costs. Some people point to these factors as justifying a different treatment for agriculture in the international economy, including justifying trade-distorting agricultural policies. This is wrong-headed; societies can support farms and preserve rural communities in ways that foster choice, protect natural resources, and expand trade.

Farm production in the next century cannot afford to be trapped in a static system in which prices are determined by government mandate, production decisions are controlled by central planners, and farmers are forced to produce only for local consumers. This myopic system cannot be sustained in any important agriculture producing society. Moreover, this type of system will not meet the needs of the coming century, when we will face unprecedented consumer demand and natural resource constraints.

Instead, I look forward to dynamic world of agricultural trade in which producers, exporters, and retailers apply the creativity of the human mind to the natural bounty of the earth. In this "new" world, we will produce a greater amount and variety of food than ever before, feed the coming billions, sustain our environment, and unlock economic resources otherwise stifled by moribund protectionism, ultimately raising living standards around the world.

Export Subsidies:

A Distortion to Free Trade in Agriculture

Export subsidies are generally considered one of the most distorting trade tools used by governments to interfere with commercial markets. Export subsidies allow a government to determine the level and direction of trade solely on the basis of government subsidies, lowering world prices and denying sales for other, more competitive exporters. Not only are export subsidies unfair commercial tools, but, by encouraging surplus production, they encourage adverse environmental practices, waste government budgets, and may delay restructuring and reform of domestic industries. Substantial progress toward eliminating export subsidies will be a critical element of the World Trade Organisation (WTO) negotiations scheduled to begin at the end of this year.

The Situation Today

Under the Uruguay Round Agreement, countries agreed to strictly limit the use of export subsidies. First, products that had not benefited from export subsidies in the past were banned from receiving them in the future. Second, where countries had provided export subsidies in the past, their future use was capped and gradually reduced over 6 to 10 years. (Developed countries were required to cut their spending on export subsidies by 36 per cent over six years while also reducing subsidised export quantities by at least 21 per cent on a commodity-specific basis.

Developing countries have until 2005 to cut spending by 24 per cent and subsidised quantities by 14 per cent). Third, countries agreed not to create new schemes that serve as disguised subsidies to get around the product-specific limits. Finally, countries recognised that export credit and food aid programmes were different and exempted them from the new budget and quantity limits, although there was agreement to negotiate disciplines on export credit programmes to ensure that they do not undermine WTO commitments.

Today, the European Union (EU) is the primary export subsidiser—accounting for nearly 85 per cent of the world total. Nearly all other countries agreed in the last round of negotiations not to use or to have only limited recourse to use export subsidies. EU farmers, responding to domestic prices that are often twice the world price, produce more products than can be consumed in Europe, but at such high prices that they can be sold abroad only with generous subsidies. These subsidies force other competitors out of the market and discourage production in countries with comparative advantage.

If the EU's extravagant domestic subsidies are the root cause of export subsidies, they are also putting serious pressure on the whole EU system. The need to impose budgetary discipline on EU farm programmes (annual cost, about $46 billion) is becoming increasingly evident, even in Europe, and the EU's goal of expanding its membership to new countries is putting pressure on it to bring its farm programmes into line with other countries, which will help reduce its need to rely on export subsidies in the future.

Areas for Resolution

The upcoming negotiations should continue the work begun in the Urguay Round and eliminate existing export subsidies. There is no economic justification for their continued use. By removing subsidized exports, world prices should increase, and farmers, particularly in the EU, will not be artificially encouraged to overproduce products that they cannot grow competitively.

In addition to eliminating export subsidies, countries should examine the rules defining export subsidies to ensure that countries do not resort to other policy tools that might allow governments to distort markets. Specially, WTO members should look closely at curbing agricultural state trading export monopolies that can exert undue market power or dispose of surplus commodities on a nonmarket basis. A recent WTO victory by the United States and New Zealand over Canada's special-class system of dairy exports shows that the existing rule against circumvention are effective but must be enforced.

Export credit and food aid programmes were addressed in the Uruguay Round agreement in recognition of the fact that those tools could be disguised as subsidies. These policies may again be on the agenda when the WTO negotiations commence next time. It will be important to ensure that the world's needy continue to have access to imported products, even when financial turmoil rolls world markets and limits the ability of developing countries to meet their food and fiber needs.

Certain large exporting nations—primarily in the EU have used export taxes as a supply management tool by intervening in the market to restrict exports when domestic stocks are low. These measures can wreak havoc in international markets, exacerbating price swings and reducing the confidence of net-food-importing countries to abandon trade barriers and rely on the international market to provide food security. Similarly, some exporting countries use differential export taxes to discourage exports of basic products (such as grains or oilseeds); they force exporters to process the product domestically (into flour or oil and meal, for example) and export the value added products.

23

The Uruguay Round and Agricultural Reform

The Uruguay Round of Multilateral Trade Negotiations (completed in 1994) continued the process of reducing trade barriers achieved in seven previous rounds of negotiations. Among the Uruguay Round's most significant accomplishments were the adoption of new rules governing agricultural trade policy, the establishment of disciplines on the use of sanitary and phytosanitary (SPS) measures, and agreement on a new process for settling trade disputes. The Uruguay Round also created the World Trade Organisation (WTO) to replace the General Agreement on Tariffs and Trade (GATT) as an institutional framework for overseeing trade negotiations and adjudicating trade disputes. Agricultural trade concerns that have come to the fore since the Uruguay Round, including the use of genetically engineered products in agricultural trade, state trading, and a large number of potential new members, illustrate the wide range of issues any new round may face.

During the past years since initial implementation of the Uruguay Round agreements, the record with respect to agriculture is mixed. The Uruguay Round's overall impact on agricultural trade can be considered positive in moving toward several key goals, including reducing agricultural export subsidies, establishing new rules for agricultural import policy, and agreeing on disciplines for sanitary and phytosanitary trade measures. The Uruguay Round Agreement on Agriculture (URAA) may also have contributed to a shift in domestic support of agriculture away from those practices with the largest potential to affect production and, therefore, to affect trade flows. However, significant reductions in most agricultural tariffs will have to await a future round of negotiations.

Tariffs, Incentives and Subsidies

Prior to Uruguay Round, trade in many agricultural products was unaffected by the tariff cuts that were made for industrial products in previous rounds. In the Uruguay Round, participating countries agreed to convert all nontariff agricultural trade barriers to tariffs (a process called "tariffication") and to reduce them. However, agricultural tariffs remain very high for some products in some countries, limiting the trade benefits to be derived from the new rules. To ensure that historical trade levels were maintained and to create some new trade opportunities where trade had been largely precluded by policies, countries instituted tariff-rate quotas. A tariff-rate quota applies a lower tariff to imports below a certain quantitative limit (quota) and permits a higher tariff on imported goods after the quota has been reached.

The Agreement on Agriculture required countries to reduce outlays on domestic policies that provide direct economic incentives to producers to increase resource use for production. All WTO member countries are meeting their commitments to reduce these outlays, and most countries reduced this type of support by more than the required amount. However, support from those domestic policies considered to have the least effect on production, such as domestic food aid, has increased from 1986-88 levels.

In the Agreement on Agriculture, 25 countries that employed export subsidies agreed to reduce the volume and value of their subsidized exports over a specified implementation period. To date, most of these countries have met their commitments, although some have found ways to circumvent them. The European Union (EU) is by far the largest user of export subsidies, accounting for 84 per cent of subsidy outlays of the 25 countries in 1995 and 1996. Despite substantial progress in reducing export subsidies, rising world grain supplies and falling world grain prices will make it difficult for some countries to meet future commitments unless they adopt policy changes.

The Uruguay Round's SPS agreement imposed disciplines on the use of measures to protect human, animal, and plant life and health from foreign pests, diseases, and contaminants. The agreement can be credited with increasing the transparency of countries SPS regulations and providing improved means for settling SPS-related trade disputes, including some important cases involving agricultural products. The agreement has also spurred regulatory reforms in some countries. The SPS agreement and the Agreement on Technical Barriers to Trade could provide a framework for disputes over genetically modified organisms (GMOs) brought to the WTO for arbitration.

Current Issues

Changes made to the multilateral dispute resolution process in the Uruguay Round may be as important to agricultural trade as the improvement in the substantive rules governing trade in agricultural goods. Initial evidence indicates that the WTO dispute settlement system is a significant improvement over its GATT predecessor. For example, a single country can no longer block the formation of a dispute resolution panel or veto an adverse ruling by blocking the adoption of a panel report. These improvements have led to a number of important agricultural trade cases being adjudicated before the WTO. The outstanding question for the WTO is whether members whose practices have been successfully challenged under the new dispute settlement procedures will live up to their obligations.

Other agriculture-related issues, including a bid for membership by a large and diverse group of potential new WTO members, the challenge of dealing with state trading enterprises (STEs) within WTO disciplines, and issues particular to developing countries, will shape the agenda for future agricultural trade liberalisation discussions. Thirty countries are currently seeking membership in the 134-member WTO. Countries seeking WTO membership accede under conditions negotiated with WTO membership through the privileged trade status with WTO member but may incur adjustment costs in reforming their trade policies and reducing tariffs to meet WTO requirements. Current WTO members gain greater access to the markets of acceding countries.

State trading enterprises, governmental and non-governmental entities that have been granted special rights or privileges through which they can influence trade, continue to be important to the trade of agricultural commodities because many countries consider them to be an appropriate means to meet domestic agricultural policy objectives. Continuing concerns about the trade practices of state trading enterprises in some WTO member countries and the potential accession of China and other countries where STEs are prominent will keep STEs on the WTO agenda.

Developing countries received special treatment in the Uruguay Round, including less stringent disciplines in reforming their trade policies than those apply to developed countries. In the next round of multilateral agricultural trade negotiations, developing countries will continue to have their own interests in the areas of special and differential treatment, export restraints, price stability, food security, food aid, and stock policies. As developing countries identify their positions, coalitions of countries with common trade interests may emerge.

24

Major Cyclones in Andhra Pradesh:

Some Observations

Cyclone is considered as the most devastating of all natural phenomena. Repeat cyclones create no sense of security to the wealth and life of the people. It is also well known that the sudden and rapidly developing cyclonic storm not only disrupts the prevailing order of life and produces danger, death and loss of property to large number of people residing within a geographical area but also creates environmental imbalances.

Andhra Pradesh is one of the states facing moderate to severe and repeated cyclones affecting the life of the people of the state and especially coastal districts. In the history of Andhra Pradesh, there is a long list of cyclones. The cyclone which hit the town of Machilipatnam in the year 1864 was one of the severest in the history of the world. The tidal wave was so vast and rushed with such force that it submerged the coastal areas deep up to twenty miles. It took many weeks for the flood water to sink. The entire area, which was submerged under seawater, became saltish and consequently infertile for many years to come. It appears that in the year 1796 also Machilipatnam and surrounding areas were hit with a big tidal wave which claimed over 20,000 people besides loss of property worth more than Rs.250 crores then. The impact of cyclone of 1864 was more on life and property of the people. More than 35,000 people and 3 lakh livestock dead and worth of nearly Rs. 600 crores property was lost. Almost all the houses at Machilipatnam were completely ruined. Even the mighty fort of Britishers turned into shambles but for one building and the Bell tower of the church.

With this calamity and Machilipatnam being prone to frequent cyclones, the Britishers shifted all their establishments to Madras, after this cyclone. The Britishers erected two monuments in memory of the 1864 cyclone victims—one of Robertson Square in the heart of the town and the other at Machilipatnam Fort.

Relief measures were taken up to persons who were deprived of the essential needs of life because of natural disaster resulting from cyclone of 1864. Cyclone relief consists largely of emergency provisions of food, clothing, medical care and shelter with the aid of voluntary contribution from other communities and countries. Even International Red Cross and other voluntary organisations did not take cyclone relief as one of the chief activities. Only in 20th century these organisations have been extending assistance to the victims besides greater role of the central and state governments in relieving the population affected by the cyclones.

The second major cyclone in the state like that of 1864 cyclone, hit the coastal region on the night of 19th and 20th November, 1977. There was terrific gale with speed ranging 120-150 Km. per hour. Coastal districts of Andhra Pradesh such as Krishna, Guntur, East Godavari, West Godavari and Prakasam and to a small extent Srikakulam, Nellore and Visakhapatnam districts were affected. The most affected areas were Divi, Machilipatnam, Bapatla, Repalle and to some extent in Chirala Taluk. Huge tidal waves engulfed the coastal region of the Divi, Machilipatam and Repalle Taluks. There was extensive damage to life and property. Approximately 71 lakh persons in 2302 villages were affected by the loss of property, crop and other damages. 7932 persons lost their lives. In Krishna district alone 6706 persons died. The loss of cattle was 2,32,046 and 45,530 other livestock. There was substantial damage to houses—86,650 huts were completely washed away. Approximately 2,16,000 persons have either lost their houses or have suffered other loss. There has been widespread damage to crops worth of Rs. 450 crores and substantial damage to public buildings such as roads, electrical installations, highways, railway, telephonic installations, irrigation canals, etc.

To overcome the effects of 1977 cyclone, 172 relief camps were opened in Krishna, Guntur, East Godavari and Prakasam districts. About 2 lakh persons were provided shelter immediately. More than 25,000 quintals of rice and food packets were distributed. Medical special staff was sent to the affected areas to assist the Collectors to take measures against out-break of epidemics. Financial assistance was given for house repairs and the Government sanctioned remission of land revenue in the most affected areas of Machilipatnam, Repalle, Bapatla and Divi taluks.

Andhra Pradesh again experienced a major cyclone disaster in May 1979. Besides loss of life there was enormous damage to public and private properties in coastal districts. Prakasam, Nellore and Kurnool districts were the worst hit. The heavy rain under the influence of the cyclone, affected the districts of Guntur, Krishna, West and East Godavari on the coastline and Kurnool and Cuddapah districts of Rayalaseema and Mahaboobnagar district of Telengana. This was most unusual occurring in the month of May and again touching all the three regions in the state. Due to precautionary measures like providing information in time to the people of the state evacuating all the people to safer places mobilising and gearing up of administrative set up to take immediate steps, death toll and loss of property were greatly reduced though the intensity of the cyclone was double than that of the one which occurred in 1977.

Nearly Rs. 60 crores worth of crops were damaged in the State. The crops in Nellore district were heavily damaged followed by East Godavari, Guntur, Cuddapah and Prakasam districts. Due to intensive precautionary measures during precyclone period, the death toll was drastically kept under limit with just 750 human lives though the severity of it was very high compared to the earlier cyclones. About 3 lakh livestock were lost due to improper care taken during pre-cyclone and during cyclone periods. Out of the six cyclone affected districts, Prakasam lost heavily its livestock. Most of the deaths in this district were due to surging waters, which breached the tanks with extensive loss of livestock and damage to the property.

More than cyclone relief measures, 1979 cyclone reveals that the success of the excellent preparedness measures taken by the State could reduce the death toll greatly. Wholesale evacuation of the people living in low lying areas in coastal areas prevented not only the number of deaths and loss of property but also could reduce the risk of undertaking relief measures.

Another severe cyclone was experienced by the people of Andhra Pradesh in October, 1983. As Andhra Pradesh is called the rice bowl of India and more especially East and West Godavari districts, paddy crop was completely damged. All the major rice producing districts such as East Godavari, West Godavari, Nizamabad and Karimnagar districts were severely affected besides Visakhapatnam, Khammam, Warrangal and other districts. Though the loss of human life was below 200, the cyclone could affect 50 lakh people in 6,322 villages of 14 districts. More than 2 lakh houses were completely damaged and nearly 1,50,000 houses were affected partially; 12,000 cattle and 31,000 livestock perished in the cyclone. Standing crops in 1,38,500 hectares were completely lost and crops in about 31,13,150 hectares were partially damaged in addition to dry crops in 3 lakh hectares of land. The loss of both public and private properties was estimated around Rs. 600 crores. The road length of more than 8,500 kilometres damaged completely. This was considered as one of the most severe cyclones in recent years as it could affect the economic structure of the State very badly.

Like that of 1979 cyclone, with utmost pre-cyclone preparation and protection, death toll and loss of property were minimised to the lowest possible level. Thanks to the immediate step taken by Telugu Desam Government and officials incharge in the districts, a major threat was averted with minimum loss by evacuating the people to safer places in almost all the coastal districts. Indian Army and Navy, Police and other officials could evacuate more than 50,000 persons in Visakhapatnam, East and West Godavari, Guntur, Krishna, Nizamabad, Karimnagar and Nalgonda districts to safer places. Hon'ble Chief Minister, Sri N.T. Rama Rao, has sanctioned immediately Rs. 30 crores towards relief without waiting for Central assistance.

After five months, in February 1984, the districts of Nellore and Chittor have again affected by a cyclone. The most unusual cyclone in the month of February could bring misery to farmers in more than 150 villages by destroying standing paddy crop and gardens worth of more than Rs. 20 crores and loss of property in the form of collapse of houses worth more than 50 crores. It is, however, the administration organised of emergency relief and rehabilitation could set right the imbalance. The Cyclone of 1996 has the same impact on A.P. economy.

The cyclones which affected the people of Andhra Pradesh demonstrate that the people affected by cyclone disaster are not confined to the immediate geographic areas of destruction, death or injury. All persons who are related to, or who identify themselves with, persons and organisations in the stricken area are also affected. Thus the effectiveness of cyclone relief measures are usually national in scope. In each cyclone period, effective preparations were made to face cyclone with necessary arrangements by evacuating the population to safe places.

But there remains a problem of co-ordination and control in undertaking relief operations. Shortly after the cyclone impact, thousands of persons have to converge on the affected area and on first-aid stations, hospitals, relief centres and communication centres near the area. Along with this movement of persons, incoming messages of anxious inquiry and offers of help from all parts of the country and even from foreign countries should come forward to set right communication facilities and to supply food, clothing, bedding and other materials. This action should continue for a few weeks following a cyclone. At times of cyclone generally there would be confusion due to lack of systematic procedures for maintaining a central strategic overview of the cyclone to apply controls and resources where they are most critically needed. Communication is often inadequate partly because of the destruction of communication facilities but more generally because of improper use of these facilities. With greater coordination and cooperation from the people and making them feel a great sense of urgency to help the victims, these effects of the cyclones would be greatly minimized in future.

Crisis Prevention:

Can Better Development Planning Lessen the Toll of Civil Emergencies and Natural Disasters?

Even a cursory scan of the world's headlines is depressing: armed conflicts are grinding on in Somalia, Afghanistan and in a growing number of other countries. And the effects of natural disasters are becoming more catastrophic each year. International relief aid, in response to such emergencies, has increased substantially. But how large can these sums of money realistically be expected to grow? With no end in sight to the need for relief, the goodwill of international donors is quickly giving way to disillusionment.

This leads us to a second question, which is, where does development fit in this grim scenario? For the development community to remain aloof from the issue of disasters and emergencies is not only politically short-sighted, it also ignores totally the causes and the effects of such phenomena.

Natural hazards such as hurricanes and earthquakes may be impossible to prevent. But they only become natural disasters if people are vulnerable. Why is it, for example, that an earthquake in Khilari, Maharashtra, that registered 6.9 on the Richter scale killed up to 35,000 people, when an earthquake of almost the same magnitude in Los Angeles in 1994 claimed only 57 lives? By reducing poverty we can help increase the coping capacity of vulnerable populations. Therefore helping people lower such vulnerability is as much a development issue as the environment, or women's participation in development.

Moreover, the repercussions of natural disasters go far beyond the immediate casualty list that so transfixes the media. Secondary and long-term effects can be equally big if not more devastating. And they must be taken into account by development practitioners.

It has been estimated, for example, that the damage to Mexico City's infrastructure from a massive 1985 earthquake amounted to US$ 3.6 billion. Yet, over the subsequent five years, the negative ripple effect on that country's balance of payments resulted in a loss of $8.6 billion. Furthermore, reconstruction requirements forced Mexican authorities to revise their economic policies to meet an increased demand for public funding, credits and imports. The priorities for public expenditure were redirected to reconstruction projects, leaving many of the pre-disaster problems of the city and its people unattended.

In Bangladesh, floods in the recent past killed 2,000 people. But on closer examination we find that the toll was much more expensive than that: in each of these years the country's economic growth rate was halved by the delayed planting of rice and the destruction of seedbeds in the floods, further undermining the country's food security. All of these are consideration that go beyond relief, but they must be taken into account by development professionals.

Other emergencies may be more complex, but must be subjected to the same analysis. As the situations in Angola, Burundi, Somalia and the former Yugoslavia demonstrate, we know little about the dynamics of emergencies that arise from civil conflict. We do know, however, that their cause usually lies in a lethal mix of poverty, poor governance and ethnic or religious rivalries exacerbated by profound social inequities. We are also learning that their resolution frequently requires the application of peacekeeping and political measures, combined with relief and development. Among the most virulent effects of such complex emergencies is the massive displacement of people; women and children are the principal victims, constituting 70 per cent of the world's refugees.

These complex emergencies around the world could easily get worse before they get better. This being said, carefully designed development efforts carried out as building blocks to national reconciliation in the fragile post-conflict stage will need to increase commensurately. The appropriateness and the sustainability of these development efforts will be one of the most important factors in determining whether peace itself becomes sustainable. For example, the absence of carefully tailored reintegration strategies for demobilised soldiers and their host communities would be an almost open invitation to resumed violence.

Yet we must also be conscious of the impact of aid and try harder to prevent the need for relief in the first place. An increasing body of evidence suggests, for examples that emergency aid can sometimes be counter-productive in the longer term, increasing the vulnerability of populations and impeding recovery. Ironically, we find ourselves in situation today where it is far easier to obtain funds for maintaining refugees in their places of asylum than for helping them reintegrate into their own societies. In such cases, we may very well be helping to perpetuate the problem that we sought to relieve, as the presence of large numbers of refugees is sometimes itself a cause of conflict.

So how are we to proceed? And what exactly is the nature of the relief to development continuum that remains logical in the abstract but elusive in reality? The concept of a continuum does not imply a linear and absolutely progressive set of responses. On the contrary, it means that we are dealing with a set of processes rather than rightly defined steps. It also means that development must be very much part of the disaster management process, and that the aim of the continuum must be to move from relief to rehabilitation and resumed development at the earliest opportunity. However, this resumed development must include conscious measures to reduce the vulnerability that caused the disaster or the emergency in the first place.

In other words, we must give greater thought to prevention before we reach for the "cure"—for humanitarian, political and financial reasons. (The Japanese insurance industry spends $200 million a year on disaster education alone). And as development practitioners, we must reconcile ourselves to the vastly more complicated environment in which we have to operate.

This means, for example, that we will have to begin examining whether the economic policy "medicine" often prescribed will reduce conflict or enhance it. We will have to ask ourselves if the reconstruction period following a civil conflict or natural disaster is the right time to advocate cuts in social spending, as has happened in certain countries in Africa and Latin America. Similarly, is it really in children's best interests to build a school in a seismic zone without first ensuring its structural stability? And does it really make sense to urge drought-prone countries to increase their reliance on cash crops, as has been done in some instances.

A story that never made headlines anywhere involves hundreds of the poorest people in Bangladesh, whose homes remained intact during the floods of 1988, when many others were simply washed away. These people were fortunate enough to have obtained credit through the Grameen Bank for construction materials as well as instruction in the building of flood-resistant homes. The Grameen revolving fund had received start-up capital from International Financial Agencies. Since that time the effort has been expanded, and more than 10,500 flood-resistant homes have been built in the last two years.

This is just one example of the kind of action we need more of—in fairly predictable and recurring circumstances such as the floods in Bangladesh—as well as in the more complex, man-made emergencies to which we must respond.

26

The Do's and Don'ts of Risk Reduction

The most effective disaster mitigation measure that can be taken at community level is for people not to build in high-risk areas such as unstable slopes, riverbeds or flood plains.

Local knowledge about these hazards is usually good, especially among older people. Sometimes, though, a hazard such as a geological fault is not obvious or visible, and surveyors and geologists have to be called in.

People can also ensure they reduce risks in their houses. Roofs should be secured against hurricanes. Roof shape is important for wind resistance. A flat roof is much more likely to be blown off than a pyramid-shaped one. Some worry about the cost of such measures, but they are no more than a small percentage of the total cost of the building and are well worth the investment.

Earthquake mitigation focuses on building codes, including correct use of steel and the strength of concrete mixes. Wooden buildings can be reinforced by braces and tying corners so as to make the structure react as a box.

Hazards in the home are not all structural. If you live in an earthquake-prone area, you should check the following:

- Are heavy objects like cabinets, TV stands and entertainment centres attached to the wall?
- Are heavy objects on the lowest shelves?
- Are water heaters and gas cylinders bolted to the wall?

- Do household members know how to turn off the gas, electricity and water supply?
- Are hanging objects such as fans and ceiling lights securely fastened?
- Are shelves fitted with wire or board to prevent objects falling off them during tremors?
- Are dangerous substances like fuel, poisons and chemicals secured against spillage?
- Are plate-glass windows and doors covered with safety film to prevent shattering?

At the national level, governments should include mitigation in their disaster management policies. Zoning and land use laws should ensure there are no buildings in an area likely to be flooded, say, once every thirty years. Golf courses and parks could be built there instead. Steep slopes would be left as wooded areas.

But planners and policy makers do not usually have this freedom. Many rivers already flow through towns, so mitigation will take the form of reducing loss of life and damage after a flood. Ground floors can be designated non-sleeping areas, levees can be built and flood warning systems and evacuation plans can be developed.

Building codes should be drafted and where they already exist, should be reviewed and strengthened. They should ensure that buildings can survive a 7 magnitude earthquake or 200 km/h winds.

Standards of Safety

Many buildings were put up before codes were drafted, so they need to be "retrofitted" by strengthening. This is more expensive than building to resistant standards. Large public buildings and bridges are prime candidates for retrofitting. But small traditional buildings should be strengthened too. Much work has been done on this in India. Chicken mesh and mortar has been effective in reinforcing walls of adobe-type buildings against earthquake damage.

Emergency facilities such as hospitals, police stations and shelters, along with water, electricity and sewage systems, must also be able to remain functional after a disaster. They should be designed to a higher standard of safety than other buildings and retrofitted where necessary. This especially applies to hospitals—crucial after a disaster—as they can be knocked out of actions without structural damage. So non-structural mitigation measures are vital.

Cost is one of the reasons cited for lack of mitigation measures in poor countries. But one could say that such countries cannot afford not to take measures. Even reducing direct damage by one per cent through mitigation would have been worthwhile. But political support for mitigation is hard to drum up because little work has been done on quantifying the benefits of mitigation as opposed to cost.

Mitigation is also very important in the natural environment. Action taken in watersheds in the mountains will eventually affect the marine environment, especially in small countries where the distance between watershed and sea is small. Eroded soil washing down into the ocean results in silting which kills off coral and fish. Coral reefs help control beach and coastal erosion, and if they are damaged the coast is more vulnerable to flooding which will in turn cause more erosion. Coral reefs are also important for fish hatching grounds, so loss of reefs will harm the fishing industry, especially in island states. Destruction of reefs and beaches also harms countries, which live off tourism. Chemical spills into waterways eventually reach the sea and also damage coral reefs and marine life. Damaged or stressed reefs are more vulnerable to natural hazards because they are less resistant to wave action.

Dialogue between environmental and disaster managers is essential. For mitigation of work, the entire society must be involved. Programmes should ensure local communities take part in planning at the same time as they ensure political and financial support at national and international level. Mitigation must be part of our daily life. If the present century "invented" mitigation and moved disaster management beyond mere response, the next must see that mitigation becomes an integral part of planning at all levels of society.

Food Security:

Availability and Access to Food

The world food situation has never been better. Enough food is being produced today that, if it were evenly distributed, no one should have to go hungry. World food production is increasing faster than population growth: per capita production increased by 5 per cent during the 1980s. Real food prices are at historic lows and have been declining for some time now. Yields of major cereals have more than doubled in the past three decades. These trends have contributed to complacency in some quarters regarding the world food situation.

Yet, more than 700 million people in the developing world do not have access to sufficient food to lead healthy and productive lives. More than 180 million children are underweight. Diseases of hunger and malnutrition are widespread. The desire to satisfy food needs has, in combination with increasing population densities and inadequate agricultural intensification, led to much degradation of environmentally fragile lands, such as forests and steep hillsides.

Over the next 20-30 years, farmers and policy makers in developing countries will be challenged to provide food at affordable prices for almost 100 million more people every year—the largest annual population increase in history. Moreover, they will have to increase food production from more productive use of the land and without further degradation of natural resources: area expansion is no longer a feasible option in most of the world.

What future food security will look like depends not on exogenous factors over which we have no control but on the decisions and actions taken by the major players: households, private and public sector agencies, governments, and the international community. If we continue to act as we have in the 1980s and early 1990s, more people will suffer from food insecurity, it will be because some or all of these players failed to act in an appropriate and timely manner.

Feeding the World: Availability and Access to Food

There is enough food in the world today to feed everyone, if it were evenly distributed. Availability of daily food energy per capita in the developing countries as a whole increased by 0.7 per cent per year during the 1980s.

Twenty-five developing countries, including about half of the African countries, were unable to assure sufficient food energy (2, 200 calories per person per day) for their populations at the end of the 1980s even if available food energy were evenly distributed within each country. This is down from 45 countries at the end of the 1970s.

However, available food is neither evenly distributed nor fully consumed. Availability of enough food at global, regional, or national levels does not necessarily mean that everyone is well fed. For people to be food secure—that is, to have access at all times to the food required for a healthy and productive life—there must be both availability of food and access to food. Access to food by households (and individuals) is conditioned by poverty: the poor usually lack adequate means to secure access to food.

Over 1.1 billion people in developing countries were living in poverty in 1993, more than 500 million in conditions of extreme poverty. South Asia is the home of about 50 per cent of the developing world's poor—more than 500 million people. Another 15 per cent are found in East Asia, 19 per cent in sub-Saharan Africa, and 10 per cent in Latin America and the Caribbean. The prevalence of poverty (the proportion of each region's population that is poor) is very high—about 50 per cent—in South Asia as well as in sub-Saharan Africa.

Today, there are more than 700 million people who do not have access to sufficient food to meet their needs for a healthy and productive life; they often go hungry adults and children also suffer from diseases associated with hunger and poverty. For almost one fifth of the total population of developing countries to be chronically hungry tarnishes the image of the world that is now considered food-secure because it produces enough food.

Great progress has been made in meeting food needs during the last 30 years. For instance, the number of underfed people declined from an estimated 976 million in 1974-76 to 786 million in late 1980s. But the problem is far from solved. Keeping up with increasing needs and demands due to population growth, income increases, and dietary changes is itself a formidable challenge.

Hunger and food insecurity have a significant effect on health and nutrition of both adults and children. They can lead to growth failure in children. About 184 million pre-school children in developing countries were underweight in 1994. About 55 per cent of these underweight children were found in South Asia and another 16 per cent in sub-Saharan Africa. The proportion of children that are underweight is higher in South Asia (almost 60 per cent), but it is also significant in sub-Saharan Africa (30 per cent) and Southeast Asia (31 per cent). It is worrisome that the number of underweight children in sub-Saharan Africa had gone up during the 1980s from 20 million to 28 million is particularly striking.

In addition to energy deficiencies, micro-nutrient deficiencies are also widespread in the developing world. About 14 million pre-school children (under the age of five years) have eye damage as a result of vitamin A deficiency. Ten million of these children are found in Southeast Asia. Between 250,000 and 500,000 pre-school children go blind each year due to vitamin A deficiency, two-thirds of these children die within months of going blind. Many more children are mildly affected. Recent research has shown that even mild deficiencies can increase mortality significantly. Vitamin A deficiencies are closely linked to diet, which can be influenced by agricultural research and policy.

Iron deficiency affects about 1 billion people in the world. particularly children and women of reproductive age. Iron deficiency leads to anaemia, which if not checked can diminish learning capacity and increase morbidity and mortality. In the developing countries, about 370 million women between 15 and 49 years of age—42 per cent of this population group—were anaemic in the 1980s. Almost one-half were in South Asia. And there are tentative indications from South Asia and sub-Saharan Africa that the prevalence of anaemia is rising in non pregnant adult women of reproductive ages.

In sub-Saharan Africa, this trend is undoubtedly associated with deterioration in general standards of living, including increased poverty and food insecurity. Anaemia partly arises from diets insufficient in iron, which again could be addressed through agricultural research and policy. For example, a possible reason why iron deficiency and anaemia are going up in South Asia may lie in the decrease in production of iron rich pulses during that same period, which in part reflects the larger research input into competing crops such as wheat in South Asia. This emphasizes the importance of considering the effects on diet and thus on health and nutrition in setting research priorities for yield-increasing research.

South Asia is the home of about half of the developing world's hungry and food-insecure people, but this population group is growing rapidly in sub-Saharan Africa. Much of the poverty and food insecurity is in rural areas, mainly in low-potential areas such as arid zones, but urban poverty is also growing rapidly.

Four Key Factors will Influence Future Food Production and Consumption

Global and regional food production and consumption during the next 10-20 years will be influenced by a large number of factors. Changes in the following four sets of factors are likely to be particularly important:

1. Economic growth and economic policies;
2. Population growth and urbanisation;

3. Rural infrastructure, agricultural production technology, and access to modern inputs; and
4. Natural resource management and environmental considerations.

The expected impact of each of these factors on future food production and consumption is considerable.

Economic Growth and Economic Policies

Economic growth must resume in the developing world, especially in sub-Saharan Africa. To support such growth, it is critical to:

- complete structural adjustment and economic reforms;
- remove external barriers to growth, such as trade distortions and subsidies in developed countries;
- liberalise trade and remove market distortions;
- enhance access by the poor to land, capital, and technology;
- expand investment in rural infrastructure, health, education, and agricultural research and technology;
- facilitate sustainability in agricultural production; and
- reverse the decline in international assistance to agriculture.

Growth in real per capita income during the 1980s was disappointing for developing countries as a whole. However, the low average rate of growth covers large variations among regions. The high rates of economic growth in Asia are expected to continue through the 1990s, while incomes in sub-Saharan Africa are expected to keep pace with population growth.

Future economic growth depends on internal policies as well as on the international policies on the environment. The extent to which current structural adjustment and economic reforms in Latin America, sub-Saharan Africa, the Commonwealth of Independent States (CIS), Eastern Europe, and selected countries in Asia and the Middle East are carried to successful completion at an appropriate speed and sequence is of paramount importance for future economic growth in those countries.

Closely related to this issue is the question of the most appropriate role of the state in a market-oriented economy with inappropriate institutions, poor infrastructure, and insufficient experience by the private sector in dealing effectively in a competitive market environment. Overreaction to past failures such as excessive and inappropriate state intervention may cause governments to take on a passive role where intervention is needed to assure that the markets function effectively and to deal with outside influences on the economy.

Future economic growth will also depend on the international trade environment, including trade distortions by developed countries, and access to external aid. Import restrictions for agricultural and non-agricultural products in Japan, the European Union, and the United States, along with domestic agricultural subsidies and implicit and explicit export subsidies for agricultural products, are of particular concern.

Population Growth and Urbanisation

If progress in economic growth is not to be undermined by rapid population growth and excessive urbanisation, effective population and migration policies are necessary to complement growth-oriented policies. Such policies must focus on:

- universal access to family planning information and technology; and
- incentives to reduce rural-urban migration, such as provision of employment in rural areas and stimulation of agricultural and non-agricultural growth in rural areas.

Although the annual growth rate is falling for the world as a whole, the population increase during the next 20-30 years, of slightly less than 100 million people a year, will be the largest ever. Approximately 97 per cent of this increase is projected to occur in the Third World, with Africa alone accounting for 34 per cent of the growth. Thus although reductions in annual population growth rates have begun to occur in Asia and Latin America, they are insufficient to counter the absolute increases. Population growth rates of these magnitudes will greatly increase the need for food and other basic necessities.

Rural Infrastructure, Agricultural Production Technology, and Access to Modern Inputs

Continued progress in all three of these areas is critical to future food security.

- Resources must be committed to infrastructure construction and maintenance. Labour-intensive public works programmes are a viable mechanism for building roads, reforesting areas, and engaging in soil conservation projects, while creating employment and income in rural areas.
- International and national agricultural research must continue to develop yield-enhancing production technology, especially in maize, millet, and other crops, as well as build tolerance or resistance in crops to pests and adverse climatic conditions.
- Farmer access to modern inputs must be facilitated through provision of credit and technical assistance. Inputs must be made available to all farmers on time and in required amounts.

The importance of investments in rural infrastructure within the context of rapid urbanisation has already been established. Even without rapid urban growth, however, such investments are needed in many developing countries, particularly the poorest ones, to facilitate agricultural and rural development. Improved rural infrastructure enhances access to export markets, modern production inputs, and consumer goods. It reduces marketing costs, promotes exchange between intracountry markets, reduces spatial and temporal price distortions, and, in general, increases efficiency in production and marketing.

However, while essential, effective rural infrastructure alone is not enough to assure agricultural and rural development and rapid increases in food production in developing countries. Yield enhancing production technology is of critical importance. Although opportunities for expansion of agricultural production into lands not currently under cultivation still exist in some countries, such opportunities are so limited that they would probably not be able to counter losses of current agricultural lands to alternative uses on a global level. Furthermore, attempts to expand agricultural production into new lands would, in most cases, require large investments in technology, tools and materials and would increase the risk of land degradation and deforestation. Thus, future increases in food production must come primarily from higher yields per unit of land rather than from land expansion.

Agricultural research has successfully developed yield-enhancing technology for the majority of crops grown in temperate zones and for several crops grown in tropical zones. The dramatic impact of agricultural research and modern technology on wheat and rice yields in Asia and Latin America since the mid-1980s is well known. Less dramatic but significant yield gains have been obtained from research and technological change in other crops, particularly maize.

Natural Resource Management and Environmental Considerations

Research, technology development, incentives, and regulations are needed to prevent environmental degradation. These measures include appropriate water management policies, reduction of subsidies that encourage wasteful use of inputs, better definition of ownership and user rights to resources including land, education of farmers to encourage appropriate use of technology and resource conservation, and the provision of alternatives to resource-degrading inputs and techniques. Since poverty is a major source of degradation, poverty eradication is justified also on environmental grounds.

The recent surge in public and private concerns about negative environmental effects of economic growth and development may, if sustained, have important implications for agricultural development and future food production and consumption. Of particular concern of the need to avoid degradation of natural resources such as land and water, as well as deforestation, water contamination, and health risks associated with the use of chemicals. Since most of the current and potential resource degradation and environmental contamination result from situations in which those who cause and possibly benefit from degradation do not pay the costs, neither the market nor the individual producers and consumers are likely to incorporate preventive measures into their behaviour. Only when sufficient damage has been done to influence significantly current or future production costs will market and producer behaviour change. The state is more likely to undertake preventive measures either through publicly funded research and technology development or through incentive policies and regulations. Extensive water-logging, salination, and associated land degradation and productivity losses resulting from inappropriate water management are of particular concern in large parts of Asia.

No Time for Complacency

Population growth will outstrip growth in food production in sub-Saharan Africa for a long time to come unless more is done to accelerate agricultural growth. Between now and 2000, the population will grow at more than 3 per cent a year, while food production is likely to grow at 2 per cent or less a year. By the year 2000, the production shortfall is estimated to increase to about 50 million tons of grain equivalent, up from the current level of about 14 million tons. The region will not have the necessary foreign exchange to import such large amounts of food. And African governments will not be able to count on enough food aid to make up the difference. If current trends continue, by the year 2020, Africa will have a food shortage of 250 million tons, which is more than 20 times the current food gap.

Poverty is expected to increase rapidly in the coming years. Sub-Saharan Africa's share of the world's poor is expected to increase from the current 19 per cent to about 28 per cent in 2000. Furthermore, the number of underweight children is expected to increase in the 1990's in sub-Saharan Africa.

Asian demand for cereals is estimated to grow at an annual rate of 2.1 per cent between now and the year 2000, where as food production is expected to grow at 1.9 per cent per year. Much of the production shortfall is likely to be dealt with through expanded imports and perhaps through expanded regional production in response to price increases.

In Latin America, by contrast, growth in food production is anticipated to exceed food demand growth: food production is estimated to grow by 3 per cent annually between 1990 and 2000, while food demand is estimated to grow by 2.5 per cent per year.

Large areas of land are rapidly being degraded and deforested. And the principal reasons for environmental degradation—poverty, high population growth, and limited access to appropriate agricultural technology—are not being dealt with effectively.

About 700 million people are food insecure, for them the food crisis has arrived. For the 10-12 million pre-school children who died in 1994 from hunger and diseases related to malnutrition, the food crisis came and went. One-third of the pre-school children of the Third World are unable to grow to their full potential and face increased risk of death and disease.

Complacency is not in order. Clearly, Malthus underestimated the power of science to expand food production. The mass starvation that was predicted for Asia in the 1970s and 1980s did not occur because science was effectively put to work to expand crop yields. However, past yield increases came about as people with foresight had made appropriate decisions. The failure to expand investments in agricultural research and technology development during the 1980s and 1990s indicates that such foresight no longer prevails. Given the long lag time between investment in agricultural research and the resulting production increases, failure to invest today will show up in production shortfalls 10 to 20 years from now. The problems associated with environmental degradation will present themselves sooner. We must not wait until a global food crisis is upon us or until the last tree has fallen to make these investments.

REFERENCES

1. FAO, FAO Production Yearbook.
2. FAO, "The State of Food and Agriculture 1992".
3. FAO, "Agriculture Towards 2010".
4. FAO, "The State of Food and Agriculture 1994".
5. FAO, Food Outlook (December 1994).
6. World Bank, World Development Report 1995.
7. World Bank, Global Economic Prospects and the Developing Countries.
8. World Food Programme, Food Aid in Review (Rome WFP 1992).
9. World Bank, Global Economic Prospects and the Developing Countries 1992. (Washington, D.C.: World Bank).

28

Food for the Billions

Will there be enough food to feed 8 billion people who will live on earth in 25 years' time? Surprisingly few people, at least in the industrial countries, seems to be overly concerned with this question. Whereas the world conferences on the environment, on women, human rights or social issues which were held in recent years were preceded and accompanied by intensive public debate, food does not seem to be a burning issue. Don't we have mountains of surplus food, people ask. Do we not have to pay our farmers to leave their land idle in order not to add to the glut on the world markets? And hasn't the Green Revolution ended famine even in countries like India which used to be a synonym for hungry people? So where is the problem?

The advance made in agricultural production since beginning against a background of imminent crisis are indeed remarkable. In only 20 years, yields of major crops like rice, maize and wheat in developing countries went up by 80 per cent, outpacing even the rapid increase in population. But this growth in yields has slowed down in recent years, and the aim of "food for all" is once again becoming elusive. About 800 million people still do not have access to enough food to meet their basic daily needs, nearly 200 million children suffer from protein and energy deficiencies, 88 countries—44 of them in Africa—have a deficit in food production.

Everyone wants to increase food security. The definition is that "food be available at all times, that all persons have means of access to it, that it be nutritionally adequate in terms of quantity, quality and variety, and that it be acceptable within the given culture". To achieve this goal, more food must be produced—much more, because we must not only adequately feed the 5.8 billion people already on earth, but also the additional two billion who will be added to world population in the next 25 years. Critics argue that the problem is not one of production alone, but one of poverty elimination. People are not hungry because there is no food, but because they have no money to buy it, these critics say. Available resources must be better distributed to end hunger in the world.

However, even if we succeed to eliminate poverty in the next few decades—a feat which appears highly unlikely—there would still be the need to boost production, because with rising incomes people also want to eat more and better food including meat. As can already be observed in the countries of East Asia, the newly acquired wealth leads to higher consumption levels which puts additional strains on available resources are getting scarcer. Agricultural lands are being degraded at alarming speed by erosion, salinity, desertification or disappear altogether due to urban or infrastructure development. It has been estimated that 40 per cent of productive land now has diminished capacity to supply benefits to humanity due to direct human impacts of land use. Water for agricultural purposes is getting scarcer almost everywhere, and there are hardly any land reserves to be brought into production to widen the agricultural base.

In this situation, there is no alternative to increasing and improving production from the existing land area. This can only be done through research which finds the best varieties which will bring the highest yields at the lowest cost to the environment. Sustainable agriculture is the key notion, one that maintains bio-diversity, uses as little chemical inputs as possible and does not over exploit water and soil resources.

In recent years, agricultural research has been neglected—partly because of the erroneous belief that with mountains of meat and lakes of milk further production increases were not desirable. Since global grain production has stagnated and world stocks have reached an alarmingly low level last year, there has been a noticeable change of mind. To raise the awareness among governments around the world that promotion of agriculture is urgent if hunger is to be avoided in the next century.

Important work is already being done by the international agricultural research institutes which promoted the Green Revolution in the sixties and seventies and are now again in the forefront of finding solutions to the daunting task of feeding 8 billion people by the year 2020. The International Rice Research Institute (IRRI) in the Philippines, the Maize and Wheat Research Institute (CIMMYT) in Mexico or institutes like ICARDA in Syria and ICRISAT in India which work on agriculture in semi-arid and dry areas, are all seeking solutions to the problem of raising production while at the same time preserving the environment. These institutions as well as national agricultural research institutions need all the support from the public and, of course, appropriate funding, to help them accomplish their task.

The scientists are optimistic that they can develop the varieties and farming systems which will allow mankind to feed everyone on earth well into the next century. But the task is not for the scientists alone. An economic and political order must also be in place which makes it possible to eradicate poverty and allow everyone to enjoy the benefits that science can offer. Feeding the billions is, therefore, not only a scientific, but first and foremost a political.

Food First

By the time this day is over, about 40,000 human beigns—mostly children—will have died from hunger, malnutrition and related causes. Today and everyday the deaths will mount, reaching an annual toll of 13 to 18 million. Few of these people will have been caught up in famine or other emergencies. Most will have suffered from a "silent" assault—the kind that seldom makes the headlines, but which claims its victims just as relentlessly.

It is intolerable that such deprivation and suffering should be allowed to exist in a world of potential food plenty. Having enough food is fundamental to all else. At the most basic level, this may entail humanitarian relief to assist people in emergency situations. In the transition from relief to development, however, we must look at systems for ensuring that societies have the capacity to produce or purchase the food they need and that it is accessible to all.

Sustainable food security fuses the goals of household food security and sustainable agriculture; it requires both. It requires looking not only at the aggregate supply of food, but also at the distribution of income and land, and at other issues: Do people have enough income to buy food? Enough land to grow their own food? Does the food distribution system deliver food where it is needed? How much food is wasted due to inadequate distribution systems? What are the implications of trends in population growth for future food needs? What is the status of women in society, and what opportunities do women have to alter rapid population growth rates? What is being done to regenerate the resource base for food production? These questions need to be asked and answered in every country.

The challenge of sustainable food security is immense, and it is growing. One billion people—20 per cent of the global population—are too poor to obtain enough food to sustain normal work. Half a billion are too poor to obtain the food needed for healthy growth of children and minimal activity of adults. Today's failure to feed people, however, may be but a prologue to a much larger failure in the future. Given likely population increases, world food output must triple over the next 50 years if the world's people are to have a nutritionally adequate diet. It will be difficult enough to achieve this expansion under favourable circumstances, and conditions may be far from favourable.

For example, according to recent estimates, an area of about 1.2 billion hectares—the size of China and India combined—has experienced moderate to extreme soil deterioration since World War II as a result of human activities. Over three-fourths of that deterioration has occurred in the developing regions from causes such as overgrazing, deforestation, land clearing, unsound agricultural practices and increased soil salinity and water-logging, largely from irrigation. Other environmental threats to the agricultural resource base include loss of water and genetic resources, adverse effects of pesticides and climate change, both local and global.

At the most aggregate level, the required increase in food production could be met if production grew at the historic average, that is, at the two per cent per annum rate achieved over the past half-century. But is this realistic? To produce three times more calories, all the land currently under cultivation around the world would, within 50 years, have to attain levels of productivity as high as those exhibited by the very best cropland today.

To this challenge add the possibility of diminished returns from the technological, energy and other inputs that have made agriculture so successful. Some experts believe that most of the potential for increased output of cereals—from improved plant varieties, from increased use of pesticides and fertilizers and from expanding the area under irrigation—has already been captured.

Viewed from this perspective, the goal of achieving sustainable food security in the decades ahead emerges as one of the greatest challenges humanity has ever faced. Agricultural output must be tripled, and people must have the income to buy the food they need. The erosion of the resource base must be halted and then reversed. Failure on any of these fronts will yield unprecedented human suffering.

What will it take to achieve sustainable food security? Obviously, the effort will have to be immense, both in size and complexity. Outlined below are a few simple (but no easy) steps that are absolutely essential elements of serious effort.

First, as citizens of the world, we must all come to see sustainable food security as a fundamental aspect of global peace and human security. This goes well beyond merely denouncing the use of food as a weapon.

Second, we must adopt concrete international goals, such as reducing world hunger by half over the next 10 years. We will never achieve the goal of sustainable food security unless we aim at specific milestones, and assess rigorously our progress in moving toward them.

Third, we must forge a true global partnership, a compact one, for sustainable food security. All countries—rich and poor—have important roles and responsibilities. There must be reciprocal responsibilities among nations, not one-way transfers.

Fourth, we must see deterioration of the agricultural resource base—terrestrial, aquatic and climatic—for what it is: a major threat to development and a major source of economic loss. Farmers are the largest group of environmental decision-makers in the world. We must ensure that they have the means to make sustainable development a reality where it counts—in the fields and fisheries.

Fifth, we must empower the people who work the land and who keep it productive. They are in the best position to decide the most appropriate ways to graft new technology onto their own traditional knowledge of seed selection, plant protection and nutrient-cycling. Special emphasis should be given to the role of women, the main providers for two-thirds of the poorest households in the developing world, as well as the producers of 60 per cent of all food grown and consumed locally.

Sixth, we must build the capacities of developing countries, both in government and in civil society. Capacity-building means empowerment for self-reliance. It means strengthening national capacities, both inside and outside government. This is essential for recognition and analysis of problems, for decision-making on courses of action and for management of systems and processes.

Seventh, not only must we build capacity in developing countries, we must also create linkages among researchers in industrial and developing countries. This will help minimise the time lag between discovery and practical utilisation. In addition, analysts from various countries must work together to examine future food security issues with different scenarios of population growth, agricultural productivity, markets and trade, climate change, loss of soil and bio-diversity and, last but not least, political instability, in order to devise options for rational choices.

We know a good deal about how to rid the world of the scourge of hunger, and how to begin to move toward sustainable food security on a global basis. We know that economic growth and prosperity are necessary, though not sufficient, conditions for eradicating hunger. We also know that development efforts must encompass not only food production, but also socio-economic factors, including sustainable livelihoods for poor families, the implications of population growth rates, the status of women and girls and so forth. We also know that good words are not enough. Now more than ever before it is crucial that we marshal the political will to achieve our goals.

30

Population Growth and Energy

It has been scarcely 200 years—the dawn of the Industrial Revolution—since humans abandoned sole reliance on firewood, other biomass fuels, and direct sunlight to meet daily energy needs. In the past half-century, global demand for energy grew twice as fast as population, as industrial nations burned coal, oil, and natural gas to fuel their economies. Over the next half-century, world energy demands are projected to continue expanding beyond population growth, as developing countries try to catch up with industrial nations.

Developing countries will see tremendous growth in energy consumption in the next half-century, as growing populations and increasing affluence combine to drive their energy demands to dizzying levels. Based on projections from the U.S. Department of Energy and the Intergovernmental Panel on Climate Change, total energy consumption in the developing world will grow by 336 per cent—nearly three times faster than population—over the next 50 years, from 3,499 million tons of oil equivalent to 15,255 million tons. By 2030, energy consumption in the developing world will likely to surpass usage in industrial nations.

Rising per capita consumption accounts for nearly two-thirds of the growth in energy demand in poorer nations, but different population trajectories can have dramatic effects on future demands. For example, assuming the same growth in per capita energy demand, moving to the low U.N. population projection will reduce total energy demands from developing countries by 2,792 million tons of oil equivalent—the output of nearly 3,000 average-sized coal-fired power plants.

In the next 50 years, the greatest growth in energy demands will come where economic activity is projected to be highest: In Asia, where consumption is expected to grow 361 per cent, though population will grow by just 50 per cent. Energy consumption in Latin America and Africa is projected to increase by 340 per cent and 326 per cent, respectively. Lower rates of population growth in Asia, compared with Latin America and Africa, mean that energy use per person will increase most in Asia. Nonetheless, in all three regions, local pressures on energy sources, ranging from forests to fossil fuel reserves to waterways, will be significant.

When per capita energy consumption is high, even a low rate of population growth can have significant effect on total energy demand. In the United States, for example, where current per capita energy demand is nearly double that in other industrial nations and over 13 times that in developing countries, the 75 million people projected to be added in the next 50 years will boost energy demands by 758 million tons of oil equivalent—roughly the same as the present energy consumption of Africa and Latin America.

World energy use per person doubled between 1950 and 1973, before confronting a short-term slowdown when restricted exports from oil-producing nations drove up energy prices. Another price shock, combined with a global economic recession, resulted in the slowdown in the early 1980s. The most recent stumbling block in energy growth followed the 1989 revolution in Eastern Europe, when energy use in the former Soviet states plummeted. Although DOE and IPCC project substantial future growth, similar forces may act to check such a development.

World oil production per person reached a high in 1979 and has since declined 23 per cent. Moreover, estimates of when global oil production will peak range from 2011 by petroconsultants to 2025 by the IPCC, signalling future price shock as long as oil remains the world's dominant fuel. Although people born in 1950 saw per capita oil production quickly double in a few short decades, those born in 2000 are likely to see it cut in half, dropping below 1950 levels.

In addition, meeting increased energy demands will require more storage and transportation infrastructure. Communities without a reliable supply of clean water or an adequate system for waste disposal may also fall short in connection to power supplies. For the estimated 2 billion who are still off the grid—and also experiencing high rates of population growth—decentralized energy technologies, such as solar roof shingles and fuel cell power generators, are likely the most feasible and affordable option for meeting increased energy demands.

Yet, it will not necessarily be the scarcity of fuel that constrains future growth in energy consumption, but rather concerns about climate change, air quality, and water quality. Growing climate concerns will require massive reductions in fossil fuel use at a time when demand for energy is soaring. A shift to renewable energy sources, such as solar energy and wind power, in addition to continued efficiency gains for power plants, cars, and appliances, holds great promise for meeting future energy demands without adverse ecological consequences.

Energy:

A Fair Deal for all

Both the supply of energy and the demand for it have spiralled in modern societies, where everyday life and changes to the environment, global as well as local, are conditioned by energy production and use. There is a crying need for a fairer share-out of material goods, energy and economic resources.

Energy comes in three forms: so-called "fossil" fuels (coal, oil and natural gas); nuclear power; and "renewable" energies (hydroelectric power, thermal or photovoltaic solar energy, wind and tide power, wood, etc.). Each of these has its own undeniable advantages and drawbacks.

Fossil Fuels

Fossil fuels are abundant and very simple to use. Oil, for example, can be very easily transported and processed, and is relatively cheap. The technology for producing its many derivatives is highly developed. What's more, it is particularly well suited for use in all forms of land, sea and air transport. Its handy fluid form and its price make it appropriate to the needs of poor communities or those that are unable to invest in capital goods.

Fossil fuels account at present for 77 per cent of all the energy produced and will, according to the most realistic projections, still account for 73 per cent in 2020. The resources will be strictly limited geographically as well as in duration, being restricted to certain regions. This state of affairs is fraught with the risk of tensions and even conflicts, owing to the strategic importance of energy supplies.

Fossil fuels are furthermore responsible for the manmade increase in the carbon dioxide content of the earth's atmosphere, with the associated danger of an increase in the greenhouse effect and, as a direct result, global warming of the order of 1° to 4°C in the next twenty years, which would adversely affect the climate and the environment. Though much uncertainty remains as to the scale of these effects, the risk is great enough to mean that every effort should be made to slow down the increasing "carbonisation" of the atmosphere due to the intensive use of fossil fuels.

Nuclear Power

The main advantage of nuclear power is that it has no effect on the carbon dioxide content of the atmosphere. As it is also cheaper (per energy unit) than hydroelectric or thermal energy, some countries have opted strongly for this way of producing electricity.

Nuclear power is, however, far from being unanimously accepted. Public opinion is very conscious of the lack of candid information and of the safety of nuclear plants, two aspects that have not always been treated, in some countries, with all the necessary care and clarity by the authorities and the operators. The public is also worried about the disposal of long-lasting radioactive wastes, an acute problem to which the experts seem confident that a long-term solution can be found. It would also be a mistake to underestimate the danger of the spread of nuclear arms, even though the main powers are now significantly reducing their arsenals of these weapons. A final point is that only those countries which can afford to make the huge investments required can put nuclear plants into operation. The investment is offset by the low cost of the fuel but is recouped only in the medium and long-term.

Renewable Energy Sources

The ecological movements, which are worried both by global warming and by the real or imagined dangers of nuclear power, would like renewable energy sources to be developed faster than is now the case. These forms of energy at present supply some 18 per cent of total demand, which puts them well ahead of nuclear power.

Technology is moving rapidly forward in this field. These forms of energy are capable of meeting the needs of communities that it would be too expensive to connect to a central grid supply. But despite improved productivity and falling costs, they remain on the whole dearer than the two previous forms. It will be a long time before they can constitute the main source of supply. Other problems that remain to be solved include the major investments required for hydroelectric power stations and the environmental damage caused by the building of dams and wind farms.

We must face the fact that as of now there is no "miracle" energy that is risk-free for humans and their environment and is also cheap and inexhaustible. There is no such thing as absolute security as regards power generation and use, and it will not be possible in the future to do without any of the above-mentioned sources. Energy demand will continue to grow as a result of irreversible technological advances, of the justified demands of the non-industrialised countries, and of population growth that is in any case set to continue for at least the next fifty years.

Some Ethical Principles

A number of imperatives must thus be borne in mind by every individual, every nation and, in particular, the citizens of the industrialised countries. These are: the right of each individual to sufficient sources of energy; our responsibility towards our children and our children's children, protection of the environment; prevention of the potential major risks from the production of energy on a massive scale; the control of costs and the need to carry on with research in all these fields.

Some of these obligations—those relating to population growth, climate change or the disposal of nuclear wastes, for example—are of a very long-term nature, while others—effort to deal with pollution caused by road transport or chemical waste disposal—are short-term. These differences of time-scale and the various possible interactions between the quantitative and qualitative aspects of the question have to be taken into account in observations of an ethical character such as the following:

- The present situation, whereby nearly one person in four in the world is without access to the energy resources he or she requires, cannot be accepted with resignation. Those with an active role in energy policy—decision-makers, industrialists, research workers and so forth—must ultimately ensure that there exist, and continue to exist, sufficient resources of sufficiently cheap energy for all countries to have access to them, regardless of their geographical or economic situation.
- There should be no pretext for unnecessarily keeping the countries of the South, which urgently need proper infrastructures, on short commons as regards energy use. This is one area where, more than in any other, people need to be informed, so that they can take part in discussion and decision-making on subjects where scientific and technological knowledge is essential.
- Our duty to future generations enjoins us to use energy resources as sparingly and rationally as possible, especially as we know that a major part of these resources may be exhausted in a century or two.
- Even though rapid progress is being made in the exploration of space, we must acknowledge the obvious fact that we have only one Earth and must therefore preserve and protect it. Since energy production and use may jeopardize our environment, there is an urgent need for appropriate measures to be taken as rapidly and as effectively as possible. The management of nuclear waste and campaigns to combat all forms of pollution arising from energy use constitute unconditional obligations in this connection.
- Whenever massive quantities of nuclear or other forms of energy are produced or transported, e.g. when oil

is transported by sea or big dams are built, major risks to life and health ensue. Absolute safety is unattainable, but the various energy authorities are nevertheless under an obligation to issue and enforce appropriate safety regulations.

- Unit cost will continue to be the main factor influencing the choice between different forms of energy. Production costs must be controlled and savings constantly sought if energy supplies are to be available to all.

- Research sometimes seems to have been neglected in work on energy production and consumption, but it is an indispensable duty. Efforts to find new sources of energy and more economical ways of using it must continue.

Further Recommendations

We must keep our eyes open for the early warning signs of potentially dangerous or irreversible situations, and react quickly to them.

Application of the "precautionary principle" remains an unconditional obligation. We must take economic and fiscal measures designed to avert the risks of tension between producers and consumers and to encourage proper control of the resource in question. The tax instrument should be used to redistribute resources between privileged and less privileged groups of the population.

The man and woman in the street and their elected representatives must be informed about everything relating to the production and consumption of every kind of energy. Parliaments should have their own scientific and technological evaluation services, as is the case in certain countries. Projections made on the basis of different economic and demographic hypotheses ought to be regularly updated.

Now it is realised that no form of energy can replace any other form, we must endeavour to preserve the balance between producers and consumers, between rich and poor and between those that are spendthrift and those that are thrifty.

32

Energy and Sustainability

Mankind's history is marked by a growing use of energy which until the end of the Industrial Revolution came largely from renewable sources. It was coal that fed the furnaces and boilers of the Industrial Revolution from the end of the seventeenth century to the nineteenth century, and drove railway transport and steamships. As well as being a useful source of mechanical energy, it was also used in the manufacture of coal gas for street lighting and in the chemical industry. In fact, coal was the principal form of energy until 1900.

Discoveries at the beginning of the nineteenth century allowed the use of electricity and revealed the relations and the interconvertibility of different forms of energy. The principles of conservation and of energy quality did not become operative until much later. Meanwhile, in 1882, the first system for producing and distributing electricity in a large city was installed. This was the beginning of the second phase in industrialisation through electrification.

Following the first successful oil drillings in 1859, Standard Oil, the first of the modern large scale oil companies, attempted the first vertical structure for overall control of the oil process. It involved extraction from the subsoil, storage, refining and final distribution. Later, the growth of derivatives, the lower extraction costs compared to coal and the greater ease and economy of transport made oil modern society's basic energy source.

The internal combustion engine led to motorisation on a massive scale by land, sea and air, and guaranteed a constantly growing market for petrol. The forties marked the start of the new petrochemical industry, which gave rise to an enormous number of new products—synthetic rubber, plastic, medicines, cosmetics, varnishes, artificial fibres, detergents, weedkillers, fertilizers, butane, propane, etc.—opening the way to the mass production of consumer goods and introducing new, non-biodegradable substances into the environment.

After World War II, ambitious programmes to produce electricity from nuclear energy were begun, in the search for a return on the enormous amounts of money invested. The economic expansion in the West during the fifties and sixties was directly related to enormous petrol consumption at a time when energy was considered plentiful and cheap. Energy consumption during these decades grew more than exponentially. The fastest developing industrial sectors were precisely the ones that consumed most energy—petrochemical industries, metallurgy, car manufacturing, domestic appliances, electricity generating, etc.—and a trend developed towards goods and services with higher energy intensity. Since 1950, increased energy production has been systematically favoured over more rational use. So much so that the increase in energy consumption has been taken as a reliable indicator of progress.

The Aftermath of the Oil Boom

The oil crises of 1973 and 1980 showed up the fragility of an energy system that was over-dependent on oil. The War in the Gulf was reminder of what was at stake for the Western economies; free access to cheap oil in the Middle East. It was therefore fear of the hardship caused by the first crisis that brought about a change in attitudes in Western countries; efforts were directed at breaking free from this dependence, diversifying supply sources, perfecting replacement energies and promoting energy-saving programmes.

The eighties marked a change in people's awareness about environmental problems. The damage was making itself felt in more and more places and eventually the global threat to our planet as a result of our energy system became clear; the composition of the atmosphere was changing and could lead to possible changes in the climate.

According to recent figures, 82 per cent of all the energy consumed in the world is produced by burning fossil fuels, 7.5 per cent from burning biomass, 5.5 per cent from the use of hydraulic energy and 5 per cent from nuclear energy. Most of our energy, in other words, is non-renewable; it runs out as we use it. As the population increases; and as it comes from fossil fuels, which on burning increase the amount of CO_2 in the atmosphere. If we add to this the accumulation of nuclear waste, the problems of access to oil deposits, constant spillages during transport and all the different imbalances involved in the world energy system, the outlook is far from sustainable.

The inequalities speak for themselves; globally, less than a quarter of the world's richest population consumes almost three quarters of the energy commercialized in the world. For example, the average annual consumption per capita in the United States is 26 times higher than in India.

The Choice of Change

Opening the way to societies that make sustainable use of energy necessarily involves increasing and improving energy efficiency, both in supply technologies and in end-use technologies, at the same time using renewable energy sources instead of fossil fuels.

Choosing the right system for the transformation of primary energy sources into energy services such as lighting, cooling, cooking, mechanical force, transport, etc., and choosing the most suitable appliances and technologies in each case is fundamental.

The truth is that a good standard of living is possible without wasting anything like as much energy. A series of relatively straightforward measures today allow a far higher level of comfort than in 1950, using one third as much energy for heating water or for washing in the home.

Petrol consumption by vehicles has dropped by 40 per cent in forty years, from 8 litres/100 kilometres to 5.3 litres in some models, and the work of improving their energy efficiency continues. In industry, the energy consumption necessary for manufacturing large intermediary products (steel, cement, paper or fertilizer) is decreasing steadily at a rate which varies between 0.5 per cent and 0.2 per cent per year according to the product.

Today's incandescent bulbs consume one twentieth as much electricity as bulbs in the twenties. The compact fluorescent bulbs now available can cut this down again to one fifth. Efficiency in lighting has increased one-hundredfold. The use of new materials and a more rational use of traditional materials allow a reduction in the amount of energy and raw materials consumed. Building a house, for example, requires 20 per cent less energy than in 1950; building a vehicle, 40 per cent less. On a global level, reducing our society's energy-intensiveness is the first step towards energy sustainability.

33

What's Driving Migration

The scale and diversity of today's migrations are beyond any previous experience. Rapid urban growth and environmental degradation in rural areas have led to internal migration affecting hundreds of millions of people. Migration is now seen as a priority issue equal in political weight to other major global challenges such as the environment, population growth and economic imbalances between regions.

Families and households form the basis for economic growth, social development and personal fulfilment. Decisions, by individual women and men on marriage—family, a place to live—shape the destinies of communities and nations. National policies and international conditions provide the context for individual decision-making. Effective development policies, including population, reproductive health and family planning policies, address this reality.

Data on national and global population trends sets the agenda for national policy. An important element of population programmes is gathering data that will allow policy-making responsive to the realities of daily life, and to the needs and aspirations of individuals.

The dominant feature of global demographics is still growth. Age distribution is a growing concern, as the numbers of young and elderly people grow, relative to the working-age population. The world is growing steadily more urban. From being a sign of strength and dynamism in the national economy, the rate and scale of urban growth has become increasingly a cause for concern. The influx of migrants to the biggest cities may be weakening both urban and rural sectors.

International migration is small in extent compared with internal movements, but has a disproportionate impact. Both internal and international migration are driven by population growth, and by inequities between countries. Migration is one of the choices which shape people's lives and the destiny of nations. But it can also be a symptom of inequity and underdevelopment. Migrants are by definition the most vulnerable members of the host community. Their living and working conditions should be protected.

Open and frank exchange of information and views between host and sending countries is needed more than ever. The aim of the international community should be to protect the right to move, but to ensure that movement is voluntary and that it stimulates rather than holds back personal and national development. "The point of departure should be the human right to live and work where one pleases, so long as it does not infringe on other people's rights to do the same."

The Urban Transformation

The rural sector is declining in importance and its contribution to national economies. It is increasingly part of a unified economy based on the city. Contact with the urban areas is easier than ever and is encouraged by rural development.

Temporary and circular migration is giving way to more permanent settlement. The largest cities are under increasing strain, and residents are encountering increasing difficulties in improving or even maintaining living conditions. Nevertheless, migration continues, driven by a variety of forces both positive and negative. The choice to move can be part of a strategy for survival or personal development; but it is often enforced by external conditions.

The urban transformation is irreversible, but the rural sectors must also be strengthened to balance the developing economy. Attention to gender issues will be crucial in ensuring a successful transition. The forces driving internal and international migration have much in common. Demographic pressures are contributing to both. As the pressures encouraging migration increase, the options for migrants become more limited. This collision is contributing to the atmosphere of crisis surrounding both urban and international migration.

Costs and Benefits

Migration is the result of individual or family decisions. But it is also part of social process. In economic terms, migration is as much a global phenomenon as trade in commodities or manufactured goods. It is part of a broader pattern, and evidence of changing economic, social and cultural relationships.

But migration may be evidence of a different kind of relationship: the combination of poverty, rapid population growth and environmental damage is a powerful destabilizing factor driving urban growth and eventually international migration. On the recipient side, migration has usually been seen as evidence of a thriving economy: today's industrial states were built in part by migrant labour, skills and investment. In today's increasingly uncertain conditions, migration may be seen as a threat to the security and well-being of the local workforce and society at large.

The only effective means to reduce migration pressures over the long term are to slow population growth; to stimulate economic growth and job creation at home, and promote the development of the individual and the family as the basic economic and social unit.

A Question of Gender

It is often assumed that most migrants are men. In reality, women make up nearly half of the international migrant population. Gender differences in social and economic roles affect migration decision making, household strategy, and the sex composition of labour migration. Attention to the gender dimension of migratory movements ought to be an important component in population and development planning.

Women frequently take the initiative in migration decisions, which may reflect limited opportunities in rural areas. Low status limits women's choices at home and may increase pressure to migrate, but it may also affect life in the host community. Opportunities may be limited by lack of education or skills, or by customer limitation on women's freedom of action outside the family or ethnic group. Paid employment for migrant women is usually in the lowest wage, least secure, and lowest status jobs, mostly in housework, child care and trade.

Most educated women end up in the same low-status, low-wage production and service jobs as unskilled female migrants. Men too experience downward mobility, but the contrast in the decline in women's employment status is far greater. Despite these disadvantages women migrants have become significant economic actors. Their status may be improved by migration, but the advantages are not clear-cut. Women's status as migrants is affected by their vulnerability, and by their lack of reproductive freedom. To ensure improved status they will need both legal protection and essential services, including reproductive health services.

Refugees

Refugees in the 1990s are overwhelmingly in Asia, Africa and Latin America. Their numbers are large, about 17 million, and growing rapidly. A further 3.5 to 4 million were thought to be in "refugee-like situations", though estimates are probably extremely conservative, and an estimated 23 million people internally displaced.

It is important to recognise the common roots of refugees and other forms of mass movement of populations. At the same time, despite the difficulty of distinguishing between political and socio-economic causes of migration, there is a clear need to distinguish between refugees and other groups of migrants. Participation in international efforts of burden-sharing would ensure that most refugee problems would be dealt within their regions of origin.

Conclusions and Policies

Migration highlights linkages and interdependencies within countries, with many implications for development agendas, including population programmes and development assistance.

Policies to regulate or moderate international migration have concentrated largely on urban growth. They have been only intermittently effective. The most successful have concentrated on stimulating rural development and the growth of alternative urban centres.

Migration is also a personal of family decision, which is affected by external conditions such as poverty or environmental degradation, improving conditions of personal and family life can make a crucial difference in the decision to migrate, reducing dependence on migration as a strategy. Because migration is the result of personal and family decisions, it can be influenced by policies that improve the quality of life.

This offers the opportunity for policies emphasizing individual development, among them education, health (including reproductive health) and family planning. Such policies are particularly relevant to the strategies must take into account gender differences in social and economic life and the differential effects of policies.

Migration decisions are about family security and long-term life-chances, rather than simply the maximisation of income. They are ultimately strategies designed to look after the individual's and the household's needs, safeguard their security, and respond to their aspirations. If the goal is to reduce migration pressures through development, it will be essential to increase the capacity but reduce the need to migrate. Long-term external support will be required to make such policies a reality, particularly in areas of rapid population growth and potential mass outward flows. Highly co-ordinated allocation of development assistance can help establish priorities and focus attention on basic needs. The challenge to both international donors and co-operating governments is to direct programme spending to the areas where it can be most effective.

Bibliography

Allchin, B. and Allchin, R. *Civilization: India and the British of Indian Pakistan Before 500 B.C.* London, 1968.

All India Economic Conference. *Economic Life of Hyderabad.* Government Central Press, Hyderabad, 1937.

Andhra Pradesh, Government of, *Hand Book of Statistics,* Andhra Pradesh, 1993-94. p. 235.

Andhra Pradesh, Government of, *Economic and Statistical Bulletin.* Vol. XXXVII, No 11. July-December, 1992. Directorate of Economics and Statistics of Hyderabad. p. 10.

Andhra Pradesh, Government of, *Hand Book of Statistics-Andhra Pradesh,* 1993-94, Directorate of Economics and Statistics, Hyderabad, 1995. p. 7.

Andhra Pradesh, Government of, *Statistical Abstract of Andhra Pradesh,* 1992. Directorate of Economics and Statistics, Hyderabad, 1994. p. 111.

Andhra Pradesh, Government of, *Statistical Abstract of Andhra Pradesh,* 1992. Directorate of Economics and Statistics, Hyderabad, 1994. p. 111.

Agro-Economic Research Centre. *'Rice in Andhra Pradesh—A study on Inter-State Variations, Kharif,* 1978, *Part-1, Report",* Andhra University, Waltair, September, 1982. pp. 15-34 and 59-69.

Apinantara, Adul. *"Cooperation and Water Conflicts Among Water Users in North Eastern Thailand Tank Irrigation".* ADC, Workshop, Bangkok, Thailand, 1981 (Mimeo).

Appadorai, A. *Economic Research Centre in Southern India* (1000-1500) A.D. University of Madras, 1936.

Arputhraj, C. *"Problems and Prospects of Tank Irrigation in Tamil Nadu".* Workshop on Modernisation of Tank Irrigation, Problems and Issues, Centre for Water Resources, Madras, 1982 (Mimeo)

Chambers, Robert, *"Men and Water: The Organisation and Operation"* in B.H. Farmer (ed). *Green Revolution,* Macmillan, London, 1977.

Cgrabheevyky, P. *Tank Irrigation and Agricultural Development*. Kanishka Publishing House, New Delhi, (India), 1992.

Coward, Sr.E. Walter, *Irrigation and Agricultural Development in Asia*. Cornell University Press, U.S.A, 1980.

Das Gupta, Kalyan Kumar, et. al. *"Problems of Management of Tank Irrigation", Impact of Tank Irrigation: A Case Study of Tumkur District*. Himalaya Publishing House, New Delhi, 1978.

Dean. D.J. and Fray. T.L. *"Discriminant Analysis of Loans for Cash Grain Farmers"*. Agricultural Economic Review, Vol. 36 (Annual) (April 1976).

Doherty, Victor S. *"A Cross Cultural Analysis of Tank Irrigation", Workshop on Modernisation of Tank Irrigation: Problems and Issues"*. Center for Water Resources, Madras, India, 1982.

Department of Agricultural Engineering. *"Development and Optimisation in the use of irrigation Water under ex-zamin Tank in Ramanathapuram District"*, Preliminary Report, Government of Tamil Nadu, Madras, India, 1982.

Elumalai, G. *"Modernisation of Tank Irrigation Systems—Farmers' Views", Workshop on Modernisation of Tank Irrigation: Problems and Issues*. Centre for Water Resources, Madras, 1982 (Mimeo)

Evaluation and Applied Research Department, Government of Tamil Nadu. *"Evaluation of Minor Irrigation Schemes"*. Second Regional Workshop on Evaluation, Madras, India, 1979.

George, P.T., Namasivayam. D. and Ramachandraiah G. *"Application of Discriminant Function in the Farmers' Repayment Performance: A study in Chingleput District, Tamil Nadu"*, Journal of Rural Development, Vol. 3 No. 3 (May 1984).

George W. Snedecor and William G. Cochran, *Statistical Methods*, New Delhi. Oxford and IBH Publishing Co., (Sixth Edition), 1967.

Gupta, S.C. *Development Banking for Rural Development*, New Delhi, Deep and Deep Publications, 1987.

Hand Book of Statistics: Anantapur District, 1991-92 & 1992-93, Chief Planning Office, Anantapur, n.d.p. IV.

Harris, D.G. *Irrigation in India*, Oxford University Press, London, 1923.

Hayami, Y., Bennagem, E. and Barker, R. *"Price Incentive Versus Irrigation Investment to Achieve Food Self-Sufficiency in Philippines"*, American Journal of Agricultural Economics, Vol. 59, No. 4, 1977, pp. 717-721.

India, Government of. Eighth Five-Year Plan, 1992-97, Vol. II. 1992. p. 56.

India, Government of, *"Report on Minor Irrigation Works in the State of Madras."* Committee on Plan Projects, 1959.

India, Government of, *Census of India 1991-Final Population Totals,* Vol. II. Series 1, Paper 1, 1992, Ministry of Home Affairs, Government of India. New Delhi. p. 115.

India, Government of, Programme Evaluation Organisation. *A Study of the Problems of Minor Irrigation,* Planning Commission Publication. No. 140. 1961, pp. 7-103.

Indian Agricultural Statistics, for the years 1891-92 to 1947-48

Janakarajan, S. *"In Search of Tanks: Some Hidden Facts"*. Economic and Political Weekly, Vol. XXVII No 26. 1993. pp. A-53—A-60

Jayabalan, B.A. *"Modernisation of Tank Irrigation in Tamil Nadu"*, Workshop on Modernisation of Tank Irrigation Problems and Issues. Centre for Water Resources. Madras, 1980.

Jha, U.M., *Importance of Irrigation in an Agrarian Economy, Irrigation and Agricultural Development.* Deep and Deep, New Delhi, 1984.

Lakadawala, D.T., *"Growth, Empolyment and Poverty"*, Presidential Address, All India Labour Economics Conference, Tirupati, December, 1977.

Madras Institute of Development Studies. *"Tank Irrigation in Tamil Nadu—Some Macro and Micro Perspectives"*. (Report-Mimeo), 1983.

Mukundan, T.M., *"The ERY System of South India,"* PPST Bulletin, Madras September, 1988.

Namerta. *"Growth and Spatial Pattern of Market Towns in Rayalaseema Region of Andhra Pradesh,"* Unpublished M.Phil. Thesis submitted to Jawaharlal Nehru University (Centre for Development Studies), New Delhi, 1989.

Narayana Rao, J.S., *"Irrigation and Planning in Andhra Pradesh"*, Conference Papers of Andhra Pradesh Economic Association, Sixth Annual Conference, 23-24 January, 1988, pp. 145-150.

National Institute of Rural Development, *Rural Development Statistics,* 1994.

Narain, Dharan, and Roy B Shyamal, *"Impact of Irrigation on Labour Availability and Multiple Cropping"*, International Food Policy Research Institute, Research Report, 2-0, 1980, pp. 7-8 and 26.

"Need for Restoration of Tanks." The Hindu, August 8th, 1991, p. 3.

Narain, et al., *"An Approach to Study of Irrigation-A Case Study of Kanyakumari District,* Economic and Political Weekly, Vol. XVII. No. 39, September 25, 1982. p. 485.

Palaniswamy, K., *"Irrigation Tank Rehabilitation"*, The New Irrigation Era, Vol XIX, No. 3, 1981, pp. 36-37.

Palaniswamy, K. and Easter, K.W., *"The Tanks of South India: A Potential for Future Expansion in Irrigation"*, Economic Report, Department of Agriculture and Applied Economics, University of Minnisote, No. ER 83.4, 1983.

Rao, G.N. and Rajasekhar, D., *"Tank Irrigation in Andhra Pradesh: A Case Study"* Conference Papers of Andhra Pradesh Economic Association, Second Annual Conference, January, 1985, pp.57-70

Rao, V.M., *"Linking Irrigation with Development—Some Policy Issues"*. Economic and Political Weekly, Vol. XII, No. 24, 1978, pp. 993-97.

Rajasekhar, D., *Land Transfer and Family Partitioning,* Oxford and IBH Publishing Company, New, Delhi and Centre for Development Studies, Trivandrum, 1988.

Rao, N.V.N. and Ram Reddy, R., *"Minor Irrigation and Tribal Development—An Empirical Study"*, Kurukshetra, Vol. XXXIV, No. 2, November 1995, pp. 31-33.

Ramaswamy, V., et al., *"Damasi : A Concept of Equity and Productivity in Irrigation"*, Wamana, Vol. V, No. 3, July, 1985. pp. 1 and 15-22.

Ramanathan, *"Tank Irrigation in Tamil Nadu"*. Unpublished M. Phil. Thesis, Jawaharlal Nehru University, 1985, pp. 70-137.

Reddy, Narasimha, D., *"A Note on the Decline of Tank Irrigation"*, Paper presented at Lokayan Workshop on Tank Irrigation, Bangalore, July 1998.

Rao, V.M., *"Linking Irrigation with Development—Some Policy Issues"*, Economic and Political Weekly, Vol. XII, No. 24, July 17, 1978, pp. 993-997.

Rami Reddy.S., *"Factors Discriminating Defaulters from Non-defaulters in Primary Credit Cooperatives"*, Indian Cooperative Review, Vol. 14. No.1 (October 1976).

Satis, S. and Sundar, A., *People's Participation and Irrigation Management: Experiences, Issues and Options,* Commonwealth Publishers, New Delhi, 1990.

Sakthivadivel, R , et al., *"A Pilot Project Study of Modernisation of Tank Irrigation: Problems and Issues"*, Centre for Water Resources, Madras, 1982. (Mimeo).

Sen Gupta, Nirmal, *Managing Common Property: Irrigation in India and the Philippines,* Sage Publications, New Delhi, 1991.

Sen Gupta, Nirmal, *"Irrigation: Traditional Vs. Modern"*, Working Paper No. 55, Madras Institute of Development Studies, Madras, April 1985, p. 5 (Mimeo).

Sen Gupta, Nirmal, *Tank Irrigation in Gangetic Bihar*, A.N.S. Institute of Social Studies, Patna, Bihar, 1982.

Sivanappan, R.K., *"Water Management in Tank Irrigation in Tamil Nadu"*, Workshop on Modernisation of Tank Irrigation—Problems and Issues, Centre for Water Resources, Madras, 1982 (Mimeo).

Sivamohan, M.V.K., and Christopher A. Scott, (Ed), *India : Irrigation Management Partnerships*, Booklinks Corporation, Hyderabad, 1994.

SPSS Inc., SPSS/PC Release 1,1 update, the Author, U.S.A., 1984, pp. B. 202-204.

Sundar, A., and Rao, P.S., *"Farmers' Participation in Tank Irrigation in Karnataka"*. Workshop on Modernisation of Tank Irrigation: Problems and Issues, Centre for Water Resources, Madras, India,1982.

Srinivasan, T.M., *Irrigation and Water Supply: South India 200 B.C.–1600 A.D.*, New Era Publications, Madras. 1991.

Subbalakshmi, V., *"Incorporation of India and Indigenous Irrigation Institutions: The case of Dasabandam in Rayalaseema"*, Conference Papers of Andhra Pradesh Economic Association, Sixth Annual Conference, January, 1988, pp. 114-128.

Smith R. Baird, *Irrigation in Southern India*, Elder & Co., London. 18567. Ch6.

Tubpin, Y., Easter, K.W., Welsch, D., *"Tank Irrigation in North-Eastern Thailand—The Returns and their Distribution"*, Economic Report, Department of Agricultural and Applied Economics, University of Minnesota, No. Er, 82-6, 1982, p. 66.

Uma Shankari, *"Tanks: Major Problems in Minor Irrigation"*, Economic and Political Weekly, Vol.XXVI, No. 39, September 28, 1991, pp. A 115-A125.

USAID, *"Thailand North-East Small Scale Irrigation"*, Washington, D.C., 1980.

Vasudeva Rao, D. *Rural Development Through Irrigation*, New Delhi, Ashish Publishing House, 1987.

Venkatram, B.R., *"Administrative Feasibility of Tank Irrigation Authority"*, Economics Program, Consultancy Report, ICRISAT, Hyderabad, Andhra Pradesh, 1980, pp. 1-4.

Von Oppen, M. and Subba Rao, K.V., *"Tank Irrigation in Semi-Arid Tropical India, Part-I: Historical Development and Spatial Distribution."* ICRISAT,' Economics Programme Progress Report, 5, Andhra Pradesh, India, 1980. p. 1.

Von Oppen, M. and Subba Rao K.V., *"Tank Irrigation in Semi-arid Tropical India, Part I, II and III"*, Progress Report, ICRISAT,Hyderabad, 1980.

Von Oppen, M. and Subba Rao, K.V., *"History and Economics of Tank Irrigation in Semi-Arid Tropical India"*. Symposium on Rain water and Dry land Agriculture, Indian National Science Academy, New Delhi, 1982, pp. 89-93.

Von Oppen, M., *"Tank Irrigation in Southern India: Adopting a Traditional Technology to Modern Socio-economic Conditions"*, Proceedings of the Consultants' Workshop on State-of-the-Art and Management Alternatives for Optimising the Productivity of SAT Alfisoils and Related Soils, ICRISAT, Patacheru, Hyderabad, Andhra Pradesh, India. 1987, pp. 89-93.

Wijayaratna, C.M., *"Uneven Distribution of Water: Causes, Consequences and Implications of System Management"*, Workshop on Water Management, Agrarian Research and Training Institute, Colombo, Sri Lanka, 1982, (Mimeo).

Yazdani, G. (Ed). *The Early History of Deccan*, Oxford University Press. London, Part-VI, 1960.

Index

T

U

V

W